Metaphysics and Scientific Realism

Eide

Foundations of Ontology

Edited by
Otávio Bueno, Javier Cumpa, John Heil, Peter Simons,
Erwin Tegtmeier, Amie Thomasson

Volume 9

Francesco F. Calemi (Ed.)
Metaphysics and Scientific Realism

―

Essays in Honour of David Malet Armstrong

DE GRUYTER

ISBN 978-3-11-057826-3
e-ISBN (PDF) 978-3-11-045591-5
e-ISBN (EPUB) 978-3-11-045500-7
ISSN 2198-1841

Library of Congress Cataloging-in-Publication Data
A CIP catalog record for this book has been applied for at the Library of Congress.

Bibliographic information published by the Deutsche Nationalbibliothek
The Deutsche Nationalbibliothek lists this publication in the Deutsche Nationalbibliografie; detailed bibliographic data are available on the Internet at http://dnb.dnb.de.

© 2016 Walter de Gruyter Inc., Boston/Berlin
This volume is text- and page-identical with the hardback published in 2016.
Printing: CPI books GmbH, Leck

♾ Printed on acid-free paper
Printed in Germany

www.degruyter.com

Contents

Francesco F. Calemi
Introduction —— 1

Matthew Tugby
Mirage Realism Revisited —— 13
1   Introduction —— 13
2   The one over many problem —— 14
3   Devitt's charge of mirage realism —— 15
4   The one over many problem and Armstrong's world of states of affairs —— 17
5   Instantiation as partial identity —— 22
6   The one over many problem and the Platonic view of universals —— 24

Francesco F. Calemi
Ostrich Nominalism or Ostrich Platonism? —— 31
1   What is it like to be an ostrich? —— 31
2   The argument from gross facts —— 31
3   The harlot argument —— 36
4   The truthmaker argument —— 39
5   Sketch for a Platonic theory of predication —— 42
6   Concluding remarks —— 47

Peter van Inwagen
In Defense of Transcendent Universals —— 51
1   Armstrong's ontological method —— 51
2   Armstrong's primary critique of transcendent realism —— 56
3   A reply to the primary critique —— 60
4   "How can distinct particulars have the same properties?" —— 64
5   Arguments, not explanations —— 68

Peter Simons
Armstrong and Tropes —— 71
1   Universals and tropes —— 72
2   Tropes and substances —— 73
3   Armstrong's objections to trope nominalism —— 74

4	Answers to the objections —— 76	
5	Some advantages of trope nominalism —— 78	
6	Remaining problems for the trope nominalist —— 80	
7	Conclusion —— 82	

Anna-Sofia Maurin
Tropes: For and Against —— 85

1	Introduction —— 85	
2	Tropes and the one over many —— 86	
3	Armstrong on what is the 'best' version of the trope view —— 88	
4	Armstrong on why there are no tropes —— 92	
5	Piling, swapping, and 'Hochberg's argument' —— 93	
6	Laws of nature and resemblance —— 96	

William F. Vallicella
Facts: An Essay in Aporetics —— 105

1	Introduction —— 105	
2	Facts as contingently existing concreta —— 106	
3	The truth-maker argument for facts —— 108	
4	Butchvarov's objections to realism about facts —— 110	
4.1	An argument from imperceivability —— 110	
4.2	An argument from impossibility of reference —— 112	
4.3	The Paradox of the Horse and the Paradox of Snow —— 113	
5	Problems with the concretist conception of facts —— 115	
5.1	The collision of the compositional and necessitarian models —— 115	
5.2	Problems with the compositionalist model —— 118	
5.3	Necessitarianism and the collapse of Armstrong's fact ontology —— 120	
6	Facts as abstract objects: Reinhardt Grossmann —— 126	
6.1	The localization argument against concrete facts —— 127	
6.2	The 'bare particular' objection to abstract facts —— 128	
7	Concluding aporetic postscript —— 129	

Javier Cumpa
Armstrong's Hidden Substantialism —— 133

1	Introduction: Is Factualism a Truth of Armstrong's Ontology? —— 133	
2	States of affairs and the problem of universals —— 134	
3	States of affairs and the problem of truth —— 135	
4	The categorial clash between factualism and the victory of particularity —— 136	

5   Concluding remarks: the ontological consequences of the clash —— 138

Kristie Miller
Persisting Particulars and their Properties —— 139
1   Introduction —— 139
2   Transdurantism —— 141
3   Objects as property bundles —— 142
4   Objects as substrata with properties —— 149
5   Location relations —— 153
6   Explanation and identity —— 157

Stephen Mumford
Armstrong on Dispositions and Laws of Nature —— 161
1   Dispositions, ontologically speaking —— 161
2   Was Armstrong's account sufficiently realist? —— 162
3   Powers, actualism and degrees of being —— 164
4   Potency and act —— 167
5   Laws to the rescue? —— 168
6   Tendencies —— 170
7   Conclusion —— 173

Andrea Borghini
Recombination for Combinatorialists —— 177
1   Introduction —— 177
2   Constituents of states of affairs —— 179
3   Recombination for combinatorialists —— 182
4   Rebutting the trickle-down objection —— 186
5   Rebutting the objection from alien possibilities —— 188
6   Conclusions —— 190

Michele Paolini Paoletti
Who's Afraid of Non-Existent Manifestations? —— 193
1   Introduction —— 193
2   The argument —— 194
3   Armstrong vs. powers —— 197
4   Getting rid of non-existent manifestations —— 200
5   Some miscellaneous concerns about Meinongianism —— 204

Tuomas E. Tahko
Armstrong on Truthmaking and Realism — 207
1   Introduction — 207
2   Truthmaking and realism — 209
3   Truthmaking as ontologically neutral — 211
4   Realism can stand on its own — 214

D. H. Mellor
From Translations to Truthmakers — 219
1   Translations — 219
2   Truth conditions — 220
3   Dispositions — 222
4   Truthmakers — 223
5   Functionalism — 225
6   Physicalism — 227
7   Beliefs and desires — 228

Francesco Orilia
Armstrong's Supervenience and Ontological Dependence — 233
1   Introduction — 233
2   The basics of Armstrong's ontology — 235
3   The no distinctness proposal — 236
4   Supervenience and ontological dependence — 238
5   Supervenience as neither sufficient nor necessary for ontological dependence — 239
6   Cases of asymmetric supervenience — 241
7   Cases of symmetric supervenience — 244
8   Instantiation and Bradley's regress — 247

Paolo Valore
Naturalism as a Background Metaphysics — 253
1   Understanding Armstrong's naturalistic position — 254
2   The under-determination of the thesis — 255
3   The negative content — 255
4   The positive content — 256
5   Is the a priori back? — 257

Index — 261

Francesco F. Calemi
Introduction

David Malet Armstrong (8 July 1926 – 13 May 2014) has been one of the most influential contemporary metaphysician working in the analytic tradition and surely the greatest 20th century Australian philosopher. His main merit is to have reestablished metaphysics as a respectable branch of philosophy placing it at the centre of the philosophical debate, and giving it the status of an authoritative and competent interlocutor of both rational and empirical sciences. By means of a rigorously argumentative approach and a sharp prose, Armstrong has built a whole metaphysical system, that is, a *comprehensive* and *unified* picture of the *fundamental structure of the world*. One of the most striking features of his metaphysics is the simplicity of its core thesis that could be recapped by Armstrong's following motto: "States of affairs rule!" (Armstrong 1989a, 43). In order to grasp the meaning of Armstrong's motto first of all one should ask what kind of entity is a *state of affairs*, and in what sense are states of affairs supposed to "rule" in a metaphysical system.

To get an answer to such questions we should first note that while Armstrong's theory of states of affairs lies at the heart of his metaphysical system, at the center of Armstrong's theory of states of affairs there is his theory of properties as universals. Armstrong's theory of universals is concerned with the so-called "problem of universals". Despite what seems to be implied by this time-honored label, there is a wide consensus among metaphysicians on the fact that, strictly speaking, *the problem of universals* is not a *single* problem, rather it is a family of woven and partially overlapping questions that traditionally has been clustered under the same heading. Broadly speaking, one can claim that the so-called problem of universals involves the existence and the nature of properties, that is, such entities as being tall, being red, having a mass, having an electric charge, or having a spin. The list that follows is representative, but not exhaustive, of the ontological, metaphysical, epistemological, linguistics, and meta-ontological issues that fall under the umbrella-term "problem of universals":

- How can two or more numerically distinct particulars to be *identical in nature* or to be of the *same type*?
- *In virtue of what* a predicate can be true of some particular entities?
- What accounts for the fact that a single predicate can be *truly applied* to a multitude of numerically different particulars?
- How can we explain *objective resemblances* occurring between particulars if not in terms of shared properties?

- What are the *referents* of the so-called abstract singular terms (e.g. "redness", "triangularity", etc.)?
- If there are properties, what are they? Are properties *non-recurrent*, viz. particular entities? Are properties *recurrent*, viz. universals?
- Are properties *necessary beings* or *contingent beings*?
- Does *property* deserve the status of a fundamental and irreducible ontological category or should it be analyzed in terms of other most fundamental ontological categories?

Since the so-called problem of universals includes all those (and many other) questions, and properties, if admitted, can be construed not only as *universals* but also as *particulars*, i.e. *tropes*, it is not quite right to talk about *the problem of universals*: one should instead use the expression "problems of *properties*". Armstrong's answer to the intensely debated problems of properties is called "a posteriori Realism", or "Scientific Realism", or "Immanent Realism" and it entails the theses that follow:

(T_1) There are genuine properties (*contra* extreme Nominalism, which denies the very existence of properties, and *contra* radical Nominalism, which aims to reduce properties to classes, or to mereological fusions, or to resemblances occurring between particulars, or to any other nominalistically acceptable entities).

Roughly speaking, Armstrong holds that there are some *phenomena* (such as sameness of type, abstract reference, objective resemblance between particulars, laws of nature, among others) that must be accounted for, and that the best explanation we can give to them involves, besides particulars, also properties construed as *primitive* entities.

(T_2) Properties are universals (*contra* Trope Nominalism, namely the doctrine that takes properties to be particulars, namely tropes).

Armstrong is rather skeptical in identifying properties with tropes, that is, in holding that properties are non-recurrent and particularized properties. Even if he shows a great appreciation towards Trope Nominalism (he writes that Trope Nominalism "can be developed as an important and quite plausible rival to a moderate Realism about universals"[1]), nonetheless Armstrong claims that his a posteriori Realism is more palatable than Trope Nominalism, in that (i) the former explains

---

[1] Armstrong 1989b, xi.

better than the latter laws of nature and resemblance, and (ii) Trope Nominalism suffers from some metaphysical problems concerning 'piling' and 'swapping' tropes.

(T$_3$) Universals are contingent and exist only "in" their bearers, not "up in the air" in a metaphorical "*topos ouranos*" (*contra* Platonism, according to which universals are necessarily existent, ontologically independent, and transcendent entities).

The rejection of the Platonist view about universals is a crucial move for Armstrong: he contends that Platonists, in putting universals in a "separate realm", cut through the unity of universals and their respective particular bearers. As a consequence of this, on the Platonist view particulars are "bloblike rather than layer-caked", that is, they lose their internal ontological structure – whereas on an Aristotelian approach, universals are "in" their bearers as genuine internal constituents. Moreover Platonism is, on Armstrong's take, a *relational theory* because on this view a particular, say a, has a property, say F-ness, if and only if a has a suitable relation to the transcendent universal F-ness. (Cf. Armstrong 1978, 64.) As such Platonism suffers from vicious regresses (as in trying to explain what it takes for a particular to instantiate a property Platonists – Armstrong says – succumb to Bradley's regress, the third man regress, and Armstrong's relation regress), and from problems concerning powers and causality (indeed Armstrong holds that, in keeping universals and particulars apart, Platonism must accept that the causal powers of a particular, depending on the properties it has, are determined by "the relations it has to the Form beyond itself"[2])

(T$_4$) Only physics can say what universals there are.

As Armstrong contends, metaphysics, on its own right, can only establish the existence and the nature of properties, and cannot tell us what universals there are. Armstrong holds that such a question can't be answered a priori, by a pure "armchair" reflection. Instead, "the best guide that we have to just what universals there are is total science" (Armstrong 1989, 87), and since "physics is the fundamental science" (*ibid.*), there are reasons – Armstrong claims – to believe that the discovery of what universals there are is a task to be accomplished a posteriori, by physics.

---

[2] Armstrong 1978, 75.

(T₅)   Universals and particulars are bound together in states of affairs (*contra* the Relational Realism that accounts for the unity of universals and their bearers by positing a *genuine relation* between them).

The notion of *state of affairs* is pivotal and pervasive to Armstrong's metaphysics. According to Armstrong ordinary particulars are states of affairs constituted by universals and bare particulars; the world as a whole is the *biggest* – hence the most comprehensive – *higher-order state of affairs* as it collects all the lower-order ones. Furthermore, strictly speaking, states of affairs are not in space-time, rather – as Armstrong holds – space-time is a *conjunction of states of affairs*. (Cf. Armstrong 1989b, 99.) In this setting laws of nature are dyadic relations of necessitation/probabilification involving universals as relata and thus laws of nature are higher-order states of affairs.[3] The modality is a function of the recombinations of the constituents of the actual states of affairs into other states of affairs.[4] Also mathematical structures, such as classes, are states of affairs.[5] The list could continue, but the point that must be stressed is that, in Armstrong's metaphysics, to say that *states of affairs rule* implies that (i) every metaphysically relevant phenomenon is to be explained in terms of states of affairs, and (ii) *everything is either a state of affairs, or exists in a state of affairs*. But what is exactly a state of affairs?

Even if Armstrong has always abided to the thesis that a states of affairs is *a particular's having a property,* or *standing in a certain relation with other particulars,* his view about the ultimate nature of states of affairs has been changed through his works. In the early stage of his career Armstrong held a "no addition of being" view about states of affairs:

> I do not think that the recognition of states of affairs involves introducing a new entity. At any rate, it seems misleading to say that there are particulars, universals and states of affairs. For it is of the essence of particulars and universals that they involve, and are only found in, states of affairs.[6]

But few years later, under the dialectical pressure of the truthmaker principle that he endorsed, Armstrong recognized that states of affairs must be something "over and above" their constituents. Let us consider a state of affairs, say a's being F: it has as constituents the particular a and the universal F. But such a state of affairs isn't just a mereological fusion of its constituents. Indeed a could exists instantiating some properties but F, the universal F could exist instantiated by some

---

3 Cf. Armstrong 1983.
4 Cf. Armstrong 1989a.
5 Cf. Armstrong 1991 and 2004, 112–124.
6 Armstrong 1978, 80.

particular but *a*: in this circumstance there would exist both *a* and *F* but there wouldn't be the state of affairs of *a's being F*. This shows, according to Armstrong, that

> *a*'s being F involves something more than *a* and F. It is no good simply adding the fundamental tie or nexus of instantiation to the sum of *a* and F. The existence of *a*, of instantiation, and of F does not amount to *a*'s being F. The something more must be *a*'s being F – and this is a state of affairs.[7]

This volume aims to celebrate Armstrong's memory bringing new understanding, and hopefully stimulating more work, on the theses just exposed (and on their implications), with the conviction that such theses constitute an invaluable heritage for contemporary research in metaphysics.

In his "Mirage realism Revised" Matthew Tugby considers the well-known and seminal exchange between Armstrong and Devitt about the *One over Many problem*, started with Devitt's paper "'Ostrich Nominalism' or 'Mirage Realism'", which accused Armstrong of seeing a problem that is not real, that is to be a "mirage" realist about universals. Tugby holds that their debate was inconclusive. Furthermore Tugby notices that, since that debate, Armstrong's theory of immanent universals has undergone a number of developments, and contends that it is necessary to reassess the One over Many problem in the light of those developments. He argues that a stronger case than ever can be made for thinking that Armstrong ought not to take the One over Many problem seriously. However – Tugby underlines – all is not lost for realists about universals who think that the One over Many problem should be taken seriously: indeed in the last part of his chapter Tugby argues that, unlike Armstrongian Realism, Platonic Realism about universals *is not* a Mirage Realism.

The first part of the present writer's "Ostrich Nominalism or Ostrich Platonism?" reviews Armstrong's major arguments against the most extreme Nominalist position, namely Ostrich Nominalism: the *argument from gross facts*, the *harlot argument*, and the *truthmaker argument*. I argue that none of them is fully satisfactory. In the second part of my work I mount a stronger case against the Ostrich Nominalism moving from a phenomenon that has been neglected by Armstrong, that is, *predicate anaphora*. Focusing on the relevance of predicate anaphora in connection with the issue of the ontological commitment conveyed by predicates, I sketch a Platonic theory of predication, putting it to work in the field of the issues raised by Armstrong's arguments against Ostrich Nominalism, and testing it with the traditional issue concerning Bradley's regress. Eventually I argue that

---

7 Armstrong 1989b, 88.

the One over Many problem is not a real problem at all, and that Armstrong's later truthmaker-driven approach to predication is to be rejected.

Peter van Inwagen's chapter "In Defense of Transcendent Universals" argues that Armstrong's version of the One over Many problem is a *non-existing problem*. As Armstrong puts it, the following question is a typical example of the One over Many problem: "What is that in virtue of which two or more things are, say, white?" But – van Inwagen says – one can give an answer to such a question only in two ways: either indicating an *efficient-cause* (e.g., two particulars, say a and b, are both white because they have been whitewashed), or indicating a *formal-cause* (e.g., two particulars, say a and b, are both white because they reflect all the component frequencies of the white light that falls upon them more or less equally). Yet, when Armstrong asks the One over Many question he has in mind a completely different type of explanation: he looks for an *ontological analysis* of the fact that a and b are both white. But to require such an explanation is, from an ontological point of view, *absurd*. Van Inwagen contends that to say – as Armstrong does – that two (or more) particulars are white because they instantiate whiteness is as absurd as to say that they are white because *it is true that they are white*. Moreover, after rejecting Armstrong's ontological methods of the inference to the best explanation, along with his constituent ontology, van Inwagen confronts his "lightweight Platonism" with the kind of Platonism criticized by Armstrong, and stresses that, given the differences between the former and the latter, Armstrong's case against Platonism doesn't undermine his version of Platonism.

Anna-Sofia Maurin's "Armstrong on Tropes" examines Armstrong's take on Trope Nominalism. In her text Maurin investigates two crucial points: Armstrong's reconstruction of what he considers the best version of the trope view, and his assessment of it. As Maurin argues Armstrong's alleged *best version* of trope theory is not the *standard* view, and above all it is not the one that exhibits the best balance cost-benefit. Besides showing that Armstrong does not present Moderate Nominalism in its strongest form, Maurin argues that Armstrong's main arguments against it (i.e. the argument from piling, the argument from swapping, the "Hochberg's argument", the argument from laws of nature, and the argument from the axioms of resemblance) fail to convince, concluding that when it comes to cost-benefit analysis, Armstrong should have been a Moderate Nominalist rather than an Immanent Realist.

Peter Simons, in his "Armstrong and Tropes", continues on the issue arguing that Armstrong's objections to Trope Nominalism can all be answered, and that a suitable trope theory is at least well placed as Armstrong's Immanent Realism. Moreover Simons pinpoints some difficulties that Trope Nominalism encounters (as every other respectable metaphysical theory). First, it seems that there could't be trope relations. Let us consider Simons's example in order to illustrate the point: the event consisting in the collision of the Titanic with a certain iceberg on the night of 14 April 1912 apparently involves two different bodies and a trope relation occurring between them. Now, if every trope is spatially located (in accord with the Trope Nominalism orthodoxy), a Trope Nominalist should assign a location to trope relations, and this seems odd. Second, there are nomic relations that appear to be something more than mere *Humean regularities*: Armstrong explains the difference between genuine nomic relations and mere regularities by appealing to laws of nature conceived as relations of necessitation (or probabilification) that have state-of-affairs-type as *relata*. This explanation is not available to the Trope Nominalist, that should instead make do with a regularist account. And this, Simons concludes, counts as a disadvantage with respect to Trope Nominalism.

William F. Vallicella's "Facts: An Essay in Aporetics" discusses Armstrong's theory of states of affairs. Vallicella presents three main theories about the nature of states of affairs: Eliminativism (in the version proposed by Butchvarov), Concretism (held by Armstrong), and Abstractism (defended by Grossmann). Vallicella's discussion moves from Armstrong's truthmaker argument for the existence of states of affairs and claims that states of affairs, whether in the Concretist conception or in the Abstractist one, turn out to be highly problematic entities under the dialectical pressure of Butchvarov's Eliminativism. Vallicella recognizes on one hand the force of the truthmaker argument for concrete facts, i.e. Armstrongean states of affairs, and on the other hand their problematic nature. If we do not admit genuine states of affairs as truthmaker for predicative sentences, then we have some truthmakerless truths; but if states of affairs are admitted as truthmakers, then they are also troublemakers. In the end Vallicella openly recognizes a provisional aporetic conclusion concerning the ontological category of *states of affairs* that could be recapped in the following dictum: *nec tecum, nec sine te*.

Javier Cumpa's "Armstrong's Hidden Substantialism" discusses Armstrong's factualism according to which the world is a world of states of affairs, focusing on what Armstrong calls the "victory of particularity," that is the claim that states of affairs are particulars. Cumpa states that in Armstrong's ontology states of affairs are designed to solve two problems: the problem of the multiple location of universals, and the problem of the ontological ground of contingent truths. Cumpa tries to show that the combination of factualism and the victory of particularity implies that Armstrong's states of affairs are *thin particulars* and that, therefore, Armstrong's ontology can solve neither the problem of the multiple location of universals nor the problem of the ontological ground of contingent truths.

Kristie Miller's chapter, "Persisting Particulars and their Properties", focuses on an argument elaborated by Armstrong, according to which *perdurantism* is preferable to *endurantism* since the former uses less metaphysical resources than the latter. Miller points out that the considerations put forward by Armstrong on the behalf of endurantism may fail to satisfy the distinguishability *desideratum*, viz., the idea that an adequate definition of endurantism must be distinguishable from those of its rival theories. The notions taken from Armstrong's philosophy are not sufficient to discriminate endurantism from *transdurantism*, according to which objects are four-dimensional entities devoid of temporal parts. In order to avoid this difficulty, Miller uses Parson's and Eagle's theories of temporal relations. These conceptual tools are sufficiently fine-grained to make sense of the difference between endurantism, perdurantism and transdurantism. Given these definitions, Miller argues that if one has to explain what grounds the identity of an object through time, one needs to complicate the endurantist metaphysics introducing several notions such as that of *haecceity*. Since these notions aren't required neither by the perdurantist nor by the transdurantist, Miller highlights that the reasons held by Armstrong against endurantism are still available. Endurantism seems to require a heavier metaphysical machinery than that of its rivals.

Stephen Mumford's "Armstrong on Dispositions and Laws of Nature" presents Armstrong's view about laws of nature and dispositions in details, and tests its strength by way of a comparative analysis with a full-blooded Aristotelian approach to *dispositions*. Armstrong holds an anti-regularist account of the laws of nature, in that he rejects any version of Humeanism on the ground that no universals generalization could ever entail a law of nature, even if laws of nature entail universal generalizations. According to Armstrong this asymmetry is best accounted for in terms of a nomic relation, i.e *necessitation*, holding between universals. Moreover Armstrong rejected an ontology of genuine dispositions because it would lead to espouse a naturalistically unacceptable Meinongian metaphysics. Instead, he proposed that the ascription of dispositions must be analyzed in terms of laws of nature that have state-of-affairs-types as relata. Mumford argues first that Armstrong's account of dispositions is not satisfactory, being based on an objectionable twofold distinction between *existent/actual* and *non-existent/non-actual* (whereas Mumford's Aristotelian view about dispositions offers a threefold distinction between what *is actual*, what *is genuinely potential*, and what *does not exist*). In holding that an object's real potency is a mere possibility, and that a mere possibility is a non-existent, Armstrong neglects that *there are* potentialities that are non-actual, and grounded in nature. Second, Mumford shows that Armstrong's theory of laws of nature is inadequate in that it analyzes *tendencies* in terms of *probabilification*, failing thus to appreciate the point that tendencies are not mere probabilities.

Andrea Borghini's chapter "Recombination for Combinatorialists" focuses on Armstrong's combinatorial theory of possibility. As Borghini states, much of the appeal of Armstrong's combinatorial theory of possibility depends on the theory's principle of recombination. Indeed, despite its centrality, the principle has received little attention in the critical literature on combinatorialism and Borghini, in his work, first sets out to discuss how to exactly formulate the principle. Then, he shows that some notable criticisms of combinatorialism are in fact criticisms to the principle of recombination that allegedly sustains the theory. Borghini focuses in particular on Sider's 'trickle down' objection concerning the recombination of mereologically complex wholes and on the objection that combinatorialism cannot account for so-called alien possibilities. The combinatorialist – Borghini contends – can rebut both objections by fine-tuning the principle of recombination.

Michele Paolini-Paoletti's, in his "Who's Afraid of Non-Existent Manifestations?", discusses the question about powers, and defends the thesis that, if there are irreducible powers such as the power to produce a certain object (*generative powers*), then there are objects that *do not exist* and they are part of the fundamental level of the universe, violating the version of the "Eleatic principle" endorsed by Armstrong ("Everything that exists makes a difference to the causal powers of something"[8]). Thus, concludes Paolini-Paoletti, generative powers come together with *Meinongianism*. After having clarified his argument, Paolini-Paoletti examines and criticizes Armstrong's attempt to reduce powers to other sorts of entities being made in his *A World of States of Affairs*. Finally, Paolini-Paoletti deals with five accounts of generative powers that are somehow alternative to Meinongianism and with some (more general) miscellaneous concerns about the truth of this doctrine.

Tuomas E. Tahko, in his "Armstrong on Truthmaking and Realism", highlights the relation holding between Armstrong's truthmaker theory and realism, and claims that while Armstrong – along with other advocates of the truthmaker theory – contends that the truthmaker theory constitutes the core of the *correspondence intuition*, there are reasons to reject such a commitment: indeed if an argument against anti-realism uses as premise the truthmaker principle (in Armstrong's formulation), an anti-realist could simply reject it on the ground that the truthmaker principle *presupposes* realism, so the former can't be used as premise in an argument for the latter. Tahko suggests that realists had better to weaken the truthmaker principle as to make it *neutral* from an ontological point of view. In this new form the truthmaker theory can help to answer one of the most cogent challenges to realism, that is, the argument that moves from the shortcomings of the correspondence theory of truth. Showing that the correspondence theory is not an essential feature of realism, and that realism can be combined with an *ontologi-*

---

[8] Armstrong 1997, 41.

cally neutral version of the truthmaker theory (conceived as a theory that provides an account of truth), the aforementioned case against realism can be successfully blocked.

D. H. Mellor's chapter, "From Translations to Truthmakers", continues on the topic of the truthmaker theory and explores it in connection with Armstrong's materialist theory of mind. Mellor argues against Armstrong that the existence of dispositions does not require a non-dispositional, or categorical, base. Moreover he puts forward a case against the identification of beliefs and desires with brain states: according to Mellor, beliefs and desires are properties possessed by *people*, while brain states are properties possessed by the *brain cells* upon which the various beliefs and desires causally depend. This is why, Mellor states, the adequate truthmakers for true statements about what we believe or desire are not brain states, but our beliefs and desires.

Francesco Orilia's chapter, "Armstrong's Supervenience and Ontological Dependence", is about Armstrong's mature ontology, the one in which states of affairs and truthmaking gain center stage. Orilia underlines that in Armstrong's mature work there is a distinctive and pervasive appeal to a certain notion of *supervenience*, on the basis of which Armstrong feels entitled to claim that the supervenient *is not additional* to the subvenient. Armstrong's way of speaking may suggest that he does not take himself to be ontologically committed to supervenient entities, namely, in his opinion, mereological aggregates, sets, properties and relations of the manifest image of the world, internal relations, kinds, dispositions, and more. Orilia's chapter explores this issue in detail, by surveying most of what Armstrong takes to be supervenient and trying to discern when it is really appropriate to view the supervenient as ontologically dependent on the subvenient. Orilia's reconstruction leads to a couple of options regarding how to conceive of the tie that brings together universals and particulars in a state of affairs, namely a *brute fact* approach and *fact infinitism*.

Paolo Valore, in his "Naturalism as a Background Metaphysics", examines both the positive and the negative content of Armstrong's naturalistic thesis, namely "the contention that the world, the totality of entities, is nothing more than the spacetime system". As Valore underlines, while the positive content of such a thesis (viz. *there is a spacetime system*) is almost trivial, its negative content (viz. *there is nothing more than the spacetime system*) constitutes an assumption that cannot be justified *a posteriori* but *a priori*. But given Armstrong's animadversion on a priori arguments, it remains unclear how one is supposed to justify the naturalism thesis. Valore concludes that in metaphysical investigation is inevitable to take as premise and background guideline some *a priori* assumption.

Bibliography

Armstrong, D. M. 1978. *Universals and Scientific Realism, Vol. 1, Nominalism and Realism*. Cambridge: Cambridge University Press.
Armstrong, D. M. 1983. *What Is a Law of Nature?* Cambridge: Cambridge University Press.
Armstrong, D. M. 1989a. *A Combinatorial Theory of Possibility*. Cambridge: Cambridge University Press.
Armstrong, D. M. 1989b. *Universals: An Opinionated Introduction*. Boulder: Westview Press.
Armstrong, D. M. 1991. "Classes Are States of Affairs". *Mind*, 100(2), 189–200.
Armstrong, D. M. 1997. *A World of States of Affairs*. Cambridge: Cambridge University Press.
Armstrong, D. M. 2004. *Truth and Truthmakers*. Cambridge: Cambridge University Press.

Matthew Tugby
Mirage Realism Revisited

1 Introduction

In the 1980 exchange between Devitt and Armstrong, Devitt accused Armstrong of being (largely) a 'mirage' realist. By this Devitt meant that Armstrong largely accepts his theory of immanent universals in order to solve a problem that is not really there. The problem in question is the one over many problem (more on this below). Even though this exchange proved to be seminal, there are reasons for thinking the main arguments on either side are inconclusive. However, since that exchange, Armstrong's theory of immanent universals has undergone a number of developments, particularly in his 1997 work, *A World of States of Affairs* and his 2004 *Truth and Truthmakers*. I think the time is right, therefore, to re-examine how Armstrong fares with the one over many problem.[1] I will argue that, where the one over many problem is concerned, there are even more difficulties for Armstrong than there were in 1980. Fortunately for Armstrong, in later work he does have other arguments for the existence of immanent universals, and so it would not be fair to think he is a mirage realist generally. Nonetheless, if the arguments of this chapter are successful, they show that Armstrong greatly overestimated the force of the one over many problem in his early work, even by his own lights. Finally, I will argue that despite our assessment of Armstrong, all may not be lost for realists who think the one over many problem is to be taken seriously. This is because the platonic view of universals, which takes universals to be transcendent entities, can serve the realist better where the one over many problem is concerned.

The structure of the chapter is as follows. In the next section I distinguish different versions of the one over many problem and identify the version that is the subject of Devitt and Armstrong's exchange. In section three I briefly outline Devitt's main reasons for thinking that Armstrong is largely a mirage realist, before examining Armstrong's response. I then suggest that there is no clear-cut winner in that 1980 exchange. In section four, I offer a new argument for the claim that

---

[1] It should be noted that Devitt has recently published a postscript to his mirage realism paper (2010, 20–30). He uses the postscript to clarify the Quinean criteria of ontological commitment and to strengthen aspects of his 1980 arguments. He does not, however, examine Armstrong's more recent views and so the arguments in this chapter are different from those offered in his postscript.

Matthew Tugby: Durham University, email: matthew.tugby@durham.ac.uk

Armstrong should not have placed any weight on the one over many problem, an argument that is based on his 1997 views. The main problem, as we will see, is that in this later work Armstrong holds that facts of the form a is F – what he calls *states of affairs* – are *ontologically basic*. On this picture, universals are merely derivative abstractions from states of affairs. I will argue that this makes it difficult to see how universals could help to explain one over many facts, given that they are ontologically dependent upon those facts. In section five I turn to Armstrong's 2004 work, in particular his new partial identity view of property instantiation. Armstrong does not say much about how this view bears upon the one over many problem, but I will argue it is far from clear that this new view of instantiation constitutes progress where the one over many problem is concerned. All in all, then, Armstrong would do best to concede that the one over many problem is not one that deserves to be answered, as Devitt urged. In the final section, however, I argue that if one abandons Armstrong's commitment to immanent universals, and instead adopts a platonic transcendent conception of universals, the prospects for giving a solution to the one over many problem are much better.

2 The one over many problem

The one over many problem was originally described by Armstrong as that of explaining 'how many different particulars can all have what appears to be the same nature' (Armstrong 1978, viii) or how distinct particulars can 'all be of the same "type"' (Armstrong, 1978, 41). There are, however, different interpretations of this problem. To begin, it is important to distinguish the semantic versions of the one over many problem from the metaphysical versions. Armstrong briefly describes the semantic problem as follows: 'it is asked how a general term can be applied to an indefinite multiplicity of particulars' (1978, xiii). In short, this version asks for a semantic analysis of *predication*, one which provides truth conditions for true predications and helps shed light on the semantic notions of designation and application. The problem of predication also has an epistemological correlate: that of how we can *know* when a predicate is being correctly applied. Indeed, one could think that these semantic and epistemological problems are genuine without thinking the metaphysical versions are, as Devitt himself urges (1980, 436). However, as Devitt highlights, it is indeed the metaphysical problem that Armstrong thinks will help to motivate realism about universals. Although one could, in principle, invoke an abundance of universals to provide the meanings for predicates, Armstrong thinks that such a view is wrong-headed (1978, xiii–xiv). What, then, is the metaphysical version of the problem? Again, there are different variants as Campbell (1990) and Oliver (1996) have highlighted. And Armstrong himself sometimes slides from one variant to another in his 1978 work. According to Campbell, the metaphysical one over many problem tends to split into two questions, which he calls the *A question* and the *B question*:

> Now we can pose two very different questions about, say, red things. We can take one single red object and ask of it: what is it about this thing in virtue of which it is red? We shall call that the A question. Secondly, we ask of any two red things: what is it about these two things in virtue of which they are both red? Let that be the B question. (Campbell 1990, 29)

Examination of Armstrong's 1978 suggests that it is primarily the B-type question that he has in mind. For instance, Armstrong's initial characterisation of the metaphysical problem, quoted above, concerns how *more than one* particular can have what appears to be the same nature. Nonetheless, Armstrong does sometimes speak of the A-type question. For example, when discussing predicate nominalism, Armstrong speaks of the need for facts of the form '*a* has the property, F' (1978, 13) to be analysed, a case involving a single particular (see Oliver, 1996, 49–50 for further examples).

Was Armstrong right to focus on both versions of the problem? Arguably not, because in the 1980 exchange, Devitt (1980, 435) points out that it is what Campbell later called the A-type question that is what Armstrong's one over many problem must really be about. What Armstrong had perhaps not noticed in earlier work is that a hard-nosed Quinean nominalist will think the answer to the B-type question is trivial from the start. Why is it that both *a* and *b* are *F*? The most obvious nominalist answer will simply point to two separate facts. The first is that *a* is *F*. The second is that *b* is *F* (Devitt 1980, 435). Surely, Armstrong must want the problem to be more difficult for the nominalist than that. Hence, Devitt concludes as follows: 'In virtue of what is *a* (or *b*) *F*? If the One over Many argument poses a problem, it is this' (1980, 435). In short, then, the one over many problem as it is set up in the 1980 exchange concerns Campbell's A-type question.

3 Devitt's charge of mirage realism

Before examining how Armstrong's later work bears upon the one over many problem, it will be useful to rehearse briefly the 1980 debate between Devitt and Armstrong. In the process, I will suggest that the debate was somewhat inconclusive.

For all Devitt says early on in favour of not taking the one over many problem seriously, he admits midway through his paper that '…if Armstrong's Realist response to the one over many argument is a genuine explanation, then there must be a problem here to be explained' (1980, 436). Hence, the most important section of Devitt's paper aims to show that Armstrong's universals offer no explanatory gain where one over many facts are concerned.

Now, as Devitt highlights, superficially Armstrong's explanation of a fact of the form a is F is that a has F-ness, where F-ness is understood as a universal (1980, 437). But as Devitt rightly points out, this is hardly illuminating unless we are told what it means for a particular to have a universal. It is at this point, Devitt thinks, that realism about immanent universals shows itself to be problematic (more on the notion of immanence in section six).

In the first instance, a realist could try construing a's having F-ness as a relational state of affairs, where the predicate 'having' represents a genuine two-place relation (Devitt, 1980, 437). The obvious problem is, however, that the one over many problem applies just as much to relational facts as it does to monadic facts. So, it looks like the relation of 'having' should be treated as a universal. But this just raises a further problem. The 'having' relation must itself be one that is *had* by its relata. And so it looks like we need to invoke a second 'having' universal to explain why the original 'having' relation holds between its relata. Clearly, this is a regress in the making. As Devitt remarks, 'this sort of Realist makes us ontologically worse off without explanatory gain' (1980, 437).

Fortunately for Armstrong, he rejects the relational view of instantiation in favour of non-relational immanent realism. But for Devitt, non-relational immanent realism is just as problematic as the relational view. According to non-relational immanent realism, the bond between a particular and the universal it exemplifies is more intimate than a relation. That is, the 'having' of a universal does not involve a further universal because the 'having' of a property does not itself involve a genuine relation. Rather, the bond between a particular and its properties is a sort of non-relational tie.

Devitt's response to this non-relational view is quite simple. He complains that Armstrong is trying 'to speak the unspeakable: to talk about "the link" between particulars and universals without saying that they are related' (1980, 437). In short, Devitt finds the notion of a non-relational link completely obscure and 'inexplicable'. Thus, he concludes that 'talk of 'particulars' and 'universals' clutters the landscape without ultimately adding to our understanding' (1980, 437).

Is Devitt's objection fair? My view is that although Devitt's objection has some force, it is by itself inconclusive. Armstrong himself concedes that the issue of how a particular stands to the universals it instantiates is 'profoundly puzzling' (1980, sect. 3). And so, he feels he is not able to say much more than that instantiation is a primitive element of the theory. Is that really so bad? At this point, I think two points are worth bearing in mind. As Devitt himself remarks, 'explanation must stop somewhere' (1980, 436). Thus, it seems Devitt does not have any objection to metaphysical primitives *per se*. The question to settle is just where the metaphysical primitives lie. In 1980 Armstrong thought that primitive 'a is F' facts are so mysterious that we need realism about universals plus a primitive notion of instantiation, whereas Devitt thought primitive instantiation is so mysterious that we should rest with primitive 'a is F' facts. Who, then, is right? I think this question cannot be answered unless we consider other relevant philosophical issues. If universals can do enough theoretical work for us elsewhere, then perhaps acceptance of primitive instantiation is worth the price. Indeed, as Armstrong points out at the end of his paper, it may be that Devitt himself has to postulate universals due to the Quinean paraphrase problem he discusses (1980, 438–9). Secondly, even if we push this first issue aside, we know in hindsight that Armstrong does have more to say about instantiation in later work. He tries to shed light on the matter by analysing instantiation in terms of the notion of partial identity (following Baxter in doing so; see Armstrong, 2004, 46–8). Of course, Devitt cannot be blamed for not addressing the partial identity proposal, because Armstrong did not formulate it until much later. Nonetheless, what this suggests is that Armstrong's notion of non-relational instantiation need not be as mysterious as Devitt supposed. Yet another reason, then, for us to reassess the Devitt-Armstrong debate in the light of more recent developments. We will return to the partial identity view in section five, where we will consider how it fares with the one over many problem.

To sum up this section, then, some of Devitt's original criticisms of Armstrong's realism are inconclusive at best. However, now that we have the benefit of knowing more about the details of Armstrong's non-relational immanent realism, we will see that those who side with Devitt can play a stronger hand. In the next section, we will say more about Armstrong's original 'one over many' explanatory strategy and see how his 1997 work on states of affairs affects it.

4 The one over many problem and Armstrong's world of states of affairs

Although Armstrong introduces the one over many problem at the start of his first book on universals (Armstrong 1978, viii), he does not always say much about the precise *form* of the explanation that his realist solution is supposed to provide. As we saw above, it is often explained that facts of the form a is F are facts in which a has F-ness, or it is sometimes said that facts of the same type share universals (e.g. Lewis 1983, 351–2). But we are not always told exactly what this sort of explanation amounts to.

However, reading between the lines, it is fairly clear that Armstrong's original idea was that talk of 'having' and 'sharing' points towards the way in which facts of the form a is F are *composed*. Indeed, in many explanatory contexts, it is entirely appropriate to explain an entity, or some feature of an entity, in terms of how that entity is *composed*. For example, if we want to know how it is that water molecules exist, an obvious explanation involves pointing out that water molecules are composed of two hydrogen atoms and one oxygen atom, and that when those three components come together in a certain way, a water molecule is created.

With this in mind, then, it seems Armstrong's original idea was that facts of the form a is F can be analysed as *composite* entities, and to say that two things share a common property is to say they have the same immanent universal as a *component*. Indeed, there are places where Armstrong explicitly uses compositional language in connection with states of affairs (see in particular Armstrong 1991).

So far, so good. However, the problem is that as soon as Armstrong got clearer on the precise nature of states of affairs (i.e., facts of the form a is F) in later work, there are reasons for thinking he undermines this explanation of one over many facts. The problem arises through two developments in Armstrong's 1997 work. The first is his insistence that the form of composition in states of affairs, if indeed 'composition' is still the right word, must be *non-mereological*. And the second development, which is closely related to the first, is that states of affairs are ontologically basic. The worry to be articulated below is that if states of affairs – like facts of the form a is F – really are *ontologically basic*, then Armstrong ultimately ends up doing just what Devitt recommends, which is to 'rest with the basic fact that a is F' (1980, 437). If this is so, then it is unclear how Armstrong, even by his own lights, can think the one over many problem is genuine. Let us now explain why in more detail.

First, why think that, if states of affairs involve composition at all, it would have to be a non-mereological form of composition? The answer lies in what some have called the 'unity' problem (see e.g. Dodd 1999). The unity problem arises from the thought that the mere existence of a particular a and property F does not by itself make it true that a is F. Assuming particulars instantiate properties contingently, then a could exist and F could exist without it being the case that a is F. The realist owes an account, therefore, of what more is needed to unify a particular and a universal within a state of affairs, as Armstrong himself emphasises (1997, 115).

One way to go, of course, is to claim that there is some further ingredient, over and above a particular and a property, which makes it the case that the particular and property are united. One way of carrying out this strategy would be to appeal precisely to the relational form of immanent realism that was discussed in the last section. This would involve appealing to a further dyadic universal which serves to bring the particular and the universal together. An advantage of this general approach is that the composition of a state of affairs would be understood in terms of garden-variety mereological composition. If the existence of the particular, the property, and the binding ingredient entails that a is F, then states of affairs can be analysed without remainder into their constituents, as all mereologically composed entities can. For this reason, Vallicella calls this approach the reductionist theory of states of affairs (2000, 238). This approach promises a straightforward and elegant explanation, but as we saw earlier, appealing to a further binding ingredient invokes further problems. For it seems we still need to explain what binds the binding ingredient to the other elements of the state of affairs. As Vallicella nicely puts it 'the existence of two boards and some glue does not entail the existence of two-boards-glued-together. Something more is needed' (2000, 244). At this point it is tempting to introduce a further binding ingredient, but as we saw during the last section in our discussion of the relational approach, this kind of remedy puts us on the road to an infinite regress.[2]

Now, we know already that Armstrong rejects the relational form of immanent realism. What this means, then, is that for Armstrong states of affairs cannot be analysed without remainder into their constituents (1997, 118). But where, we might ask, does this leave his view about the composition of states of affairs? The answer is that he can only say it is states of affairs themselves, rather than a binding ingredient, that holds the constituents together. States of affairs come with a ready-made unity, so to speak, and 'hold their constituents together in a non-mereological form of composition' (1997, 118). This non-mereological view is what Vallicella calls the *non-reductionist* view of states of affairs (2000, 246). Crucially, this non-reductionism then quickly leads Armstrong to declare that 'it is often convenient to talk about instantiation, but states of affairs *come first*' (1997, 118; my emphasis). And when Armstrong says that states of affairs *come first*, this marks a significant development in his view. It is unified states of affairs, i.e. facts of the form a is F, that are the fundamental building blocks of reality on this picture. The world is, as the title of Armstrong's 1997 book says, nothing more than *A World of States of Affairs*.

Where does all of this leave universals and particulars? The answer is that insofar as we can talk about particulars and universals, this is only because they

---

[2] For the purposes of this chapter, I have offered only a brief characterisation of this regress problem, one which is commonly attributed to F. H. Bradley. For further details of the problem and of how various versions of the mereological approach arguably fail, see Vallicella 2000.

can be 'abstracted' from a prior state of affairs. More precisely, a universal is 'everything that is left in the state of affairs after the particular particulars involved in the state of affairs have been abstracted away in thought' (1997, 29).[3] In short, without states of affairs, particulars and universals are nothing. Using a theological metaphor, we could express this picture as follows: Once God has decided which contingent properties particulars are to have, He does not gather the components (i.e., particulars and universals) and 'build them up' from there. Rather, he has to create a state of affairs *in its unified entirety, in one fell swoop*. Insofar as particulars and universals can be taken as entities in their own right, this is only because they can be abstracted from a prior state of affairs. And so, the sorts of compositional explanation that are available in, say, the molecule case discussed earlier, do not apply in the case of states of affairs, since the direction of ontological priority is different.

Where, then, does this development leave Armstrong's original 'one over many' project of providing a metaphysical analysis of one over many facts – facts of the form *a* is *F*? The difficulty is clear. Armstrong's 1997 work involves moving away from the idea that the components of a state of affairs are ontologically prior. States of affairs are now viewed as primitive unities that, in Armstrong's words, 'come first', with the universals being wholly 'dependent' on them (Armstrong 1989, 43).[4] This gives rise to the following explanatory puzzle, however. It is precisely these states of affairs – facts of the form *a* is *F* – that the one over many problem invites us to explain. But if states of affairs are ontologically basic, surely there is nothing we can be expected to explain. Indeed, if there is anything to explain, surely it will be the universals themselves, given that states of affairs 'come first' and universals are derivative abstractions from them. The problem here is that, generally, explanatory relations in metaphysics run from dependee

---

[3] It is interesting to note that the seeds of this idea are arguably present in Armstrong's 1980 exchange with Devitt when Armstrong describes particulars and universals not as entities in their own right but rather as two 'factors' (1980, sect. 3) of a state of affairs. Talk of 'factors' is suggestive of a view that takes particulars and universals to be different aspects of what is a more basic, unified entity.

[4] Mumford thinks that Armstrong's 'states of affairs first' view may even compromise Armstrong's realism about universals. Mumford writes: 'Armstrong really does have a states of affairs ontology. Such Tractarian facts are what do all the work. Universals and particulars both then seem to fall foul of the naturalist commitment'. He goes on: 'Both abstraction and partial consideration sound too much like mind-dependence. If one really is to hold a states of affairs ontology, it might be that universals and particulars will have to be sacrificed' (2007, 105). Even though I think this criticism is to be taken seriously, we will not pursue it here. The key point for the argument to follow is that *even if* universals can still be said to be real in some sense, their explanatory roles are limited given that they are no longer viewed as fundamental ontological posits.

to depender: if A metaphysically explains B then this explanatory relation will typically be grounded in the fact that B derives from A rather than vice versa (see e.g. Fine 1982 and Correia 2008). To use a common example from Fine, it is held that the singleton set of Socrates is explained by the existence of Socrates because the singleton set is ontologically (i.e., essentially) dependent on Socrates (but not vice versa: see Fine, 1995). In short, then, positing explanatory relations which run from the dependers to the dependees is to posit a relation running in the wrong direction. Perhaps a case can be made for states of affairs being an exception to the rule, but it is far from clear how such an argument would run, and Armstrong himself does not provide one. Indeed, Armstrong says surprisingly little about the one over many problem in his 1997 and 2004 work, preferring instead to focus on other arguments for universals such as the argument from laws (1997, Ch. 15) and the truthmaking argument (1997, 113).[5] This might make one wonder whether Armstrong himself came to realise that, even by his own lights, the one over many problem is a red herring.

In short, the discussion above suggests that if states of affairs – facts of the form a is F – are the basic units of reality, then the best thing to say is simply that there is not a one over many problem to be answered. Once we get to facts of the form a is F, i.e. states of affairs, we have hit the bedrock of reality. As Devitt puts it, 'we should rest with the basic fact that a is F' (1980, 437).[6] There may be good reasons for believing in Armstrong's universals, but the one over many problem should not be counted among them. To think otherwise is to be, at least in part, a mirage realist.

Finally, was Armstrong aware of these issues at the time of the 1980 debate with Devitt? Perhaps not, but there is evidence to suggest that elements of the picture above were bubbling under the surface. Interestingly, when summarising Armstrong's response to the problem of unity, Dodd includes a short Armstrong quote that is precisely from his 1980 exchange with Devitt:

> Here is what Armstrong has by way of an alternative approach. States of affairs are ontologically basic unities, their constituents, particulars and universals, being 'vicious abstractions' from them (Armstrong, 1980). Hence, the problem of how particular and universal can

---

5 In his postscript, Devitt also rightly points out that the one over many argument becomes less important in Armstrong's later work, given his new realist strategy of providing truthmakers which make our scientific theories true. Devitt says of the truthmaking strategy that 'Whatever justified that supposition would establish Realism about states of affairs without appeal to the One Over Many argument' (2010, 24, fn. 13).

6 This is not to say that there are not other problems with Armstrong's states of affairs. For recent criticisms, see Bynoe (2011) and Rissler (2006). Unfortunately, the issues discussed there fall beyond the scope of this chapter.

be unified in a state of affairs rests on a misunderstanding; it is the product of wrongly taking particulars and universals to be ontologically prior to states of affairs (1999, 151; format of Armstrong reference brought in line with the formatting of this chapter).

Nonetheless, Devitt can hardly be blamed for not making use of the above arguments, because Armstrong's view on states of affairs did not become fully clear until his 1997 work. Had Devitt known about these developments, he would have been able to play a stronger hand and argue that *even by Armstrong's own lights*, the one over many problem should not be considered a genuine one.

However, before resting with all of these conclusions, it is worth bearing in mind that in his 2004 work, Armstrong's view on universals and states of affairs underwent another significant change. It is well worth considering, therefore, whether these late developments allow Armstrong to approach the one over many problem in a different way.

5 Instantiation as partial identity

In the last section we focused on the problem of unity in order to bring out certain features of Armstrong's later view on states of affairs. An implicit assumption of that discussion was that particulars instantiate at least some universals *contingently*. For if particulars could not fail to instantiate the properties they have, then clearly the question of why certain particulars and certain universals are tied together would not be interesting. The answer would just be that things could not be otherwise. In recent work, however, Armstrong (2004, 2005a, 2005b and 2006) has had a change of heart and now views the instantiation of immanent universals by particulars as a necessary rather than contingent matter. This presents us with a rather different metaphysical view of instantiation to that discussed thus far and so our discussion would not be complete without taking it into consideration.

Following a view put forward by Baxter (2001), the source of the necessity of instantiation is said to lie in the fact that there is partial identity between a particular and each of the universals it exemplifies. Baxter holds that the link between a particular and its universals is still contingent because he thinks there can be contingent identities. Armstrong, however, has long held that identities hold necessarily, a view which he naturally carries over to the notion of partial identity.

How are we to make sense of this partial identity? Armstrong describes the partial identity of some particular a with property F as an 'intersection with F' (2004, 47). This talk of 'intersecting' indicates that instantiation involves a sort of overlap, which is to say that part of the particular and part of the universal coincide, in the same way (to use one of Armstrong's examples of partial identity) that two adjoining terrace houses have a coinciding wall (Armstrong 1997, 18). Such overlap is surely a symmetric affair: if a overlaps with b, b overlaps with a. Because of this symmetry, the relationship between particulars and universals is given a new spin. Realists about universals have long described a universal as a one that runs through many particulars. But equally, Armstrong (2006) now says, a particular is a one running through many properties.

Where, then, does this new view leave us in respect of answering the one over many problem? This is not entirely clear. As far as I know, Armstrong did not explicitly address the one over many problem in light of his new theory. However, I will now put forward reasons for thinking that this new view is ill-suited to providing a very illuminating explanation for one over many facts.

The problem is this. A consequence of Armstrong's new theory is that universals cannot be said to be wholly present in each of their various instantiations. Given the necessity of partial identity, if any particular were to have different properties to those it actually has, then it would be a different particular. But given that overlap is plausibly symmetrical, what is equally true on Armstrong's new view is that if any worldly particular were to have had different properties, the world would not have contained the very same universals. What we would have, instead, are universals some of which are merely *close counterparts* of the actual universals. In Armstrong's words:

> Consider first the particular and its properties. Could the particular have lacked any property that it in fact has? Strictly, no. Necessarily, the particular would have been at least a little different from what it actually is, and therefore would not be the same particular ... So if the particular is supposed to lack that property, will not the universal be a different entity? I think it must be. Having just the instances it has is essential to the universal being what it is. So the particular must have that property. (Armstrong 2005b, 318).

We can now clearly see why universals are not wholly present in each of their instantiations on this view. If universals were wholly present in their instantiations, then it would not be possible to take a universal out of existence by taking one of its particular instantiations out of existence (assuming it had more than one instantiation). On Armstrong's new view, however, this clearly is possible, as the above quotation implies. What Armstrong has done, then, is to shift his notion of instantiation towards the Platonic notion of *participation*, as Armstrong readily admits in one place (2006; see also Mumford, 2007, 187, where the comparison with Plato's participation view is drawn).[7]

---

[7] It is worth pointing out that, in another place, Armstrong (2005c, 274) considers moving away from the participation view and flirts with the idea that it is not essential to a universal that it

Why, then, might this be problematic with respect to the one over many problem? Well, if instantiation is a form of participation, then the initial answer we would get to the one over many problem is that a is F because a overlaps with part of universal F, while b is F because it also overlaps with part of universal F. Is this explanation satisfactory? On its own, I do not think so. The immediate question that this answer invites is *what makes it the case that the overlapping parts in question are parts of the same universal*? In other words, what unifies these overlapping parts under the same universal? Until this question is answered, we have not dealt with the point that the one over many problem is raising. Pointing out that a's nature and b's nature are overlapping parts of the same universal could be the beginning of an answer to the problem, but only if we can then explain what it means for two distinct natures to be parts *of* the same universal. What is needed, in short, is some principle of unity. And it is far from clear that Armstrong's new view can provide one. As far as I can tell, it is simply left as a brute fact that certain overlapping parts are parts of the same 'universal' and, indeed, that the parts in question are parts *of* a universal at all.

Ironically, the problem just discussed is precisely that which Armstrong once raised in his early work against the participation interpretation of platonic realism. In the quotation below, where Armstrong writes 'Form', we could replace that word with 'immanent universal' and the worry expressed would then apply to Armstrong's own partial identity view:

> If "participation" is understood literally, then each particular simply gets a numerically different part of the Form. This is clearly unsatisfactory. The problem is to explain how different particulars can all have something in common. But if the Form has to be broken up among the particulars, then the problem of what is common to the particulars is replaced by the problem by what criterion these parts of the Form are all accounted parts of the same thing. Which is no advance at all (Armstrong 1978, 66).

6 The one over many problem and the Platonic view of universals

In this final section I want to suggest that all may not be lost for realists about universals who think the one over many problem is real and who want a genuine

---

has just the instances it has. But as Mumford rightly points out, it is hard to see how this can consistently be maintained, because it is overwhelmingly plausible that partial identity is, like identity, symmetrical (2007, 191). Hence, if the partial identity view is to be maintained, it is hard to see how something like the participation view can be avoided.

solution to it. Platonic realism is, I suspect, better suited to solving the one over many problem, providing it moves away from the participation view of instantiation. Note that the discussion so far has focused firmly on the Armstrongian theory of universals, which as we have pointed out at various places is a *non-relational immanent realist view*. Perhaps if we reject the immanentist component, a more promising answer to the one over many problem can be found. But first, what precisely does it mean to say that universals are immanent? Typically, the view is characterised in terms of what Armstrong calls the *principle of instantiation*. Armstrong first expressed the principle as follows:

Principle of Instantiation: For each N-adic universal, U, there exist at least N particulars such that they U (Armstrong 1978, 113).

The principle is, I think, intended to have modal force. The immanent realist does not mean to say that, *as it happens*, each universal that exists is instantiated. Such a claim is entirely compatible with the opposing view of platonism, for it would be a mistake to think that, by definition, platonists must hold that there are uninstantiated properties. For the platonists, universals exist independently of their concrete instantiations, but whether or not all universals happen to be instantiated in our world is determined by which universals there are and what the concrete world contains. What this suggests is that the principle of instantiation is really a claim about *generic ontological dependence*. That is, if a universal is immanent then *necessarily*, if it exists, then there is at least one instantiation of it.

There is more one could say about the principle of instantiation, but for the purposes of this section I want to focus on the realist rival to the immanent view. As indicated above, platonism rejects the principle of instantiation and this raises the question of whether platonism, as an alternative to the immanent view, can do better where the one over many problem is concerned. I suspect it can.

Before addressing the one over many problem from the perspective of platonism, we first need to establish whether there is anything more to be said about platonism other than that it rejects the principle of instantiation. Clearly, the mere denial of the principle of instantiation leaves it open as to whether platonic universals are ontologically prior to worldly states of affairs or whether each category of entity is equally fundamental, existing independently of the other. Now, it is perhaps not an incoherent view to suppose that worldly states of affairs and transcendent universals 'float free', bearing no dependence relations in either direction. But it is hard to see what the theoretical point of positing universals would be in that case. Plato himself clearly thought that the 'Forms' (i.e., the transcendent universals) were more fundamental than concrete states of affairs, and that worldly entities were mere imperfect shadows cast by the perfect Forms. In the *Phaedo* (1975, 95–106) there is even the suggestion that worldly states of affairs are *causally* dependent on universals in some sense. The causal relation does not sound like the right sort of relation to modern ears in this context, but the general idea that worldly states of affairs are dependent on universals can be cashed out in other ways.

Indeed, I take it that a promising way of cashing out the relationship between platonic universals and their concrete instances is to do so in terms of the notion of ontological dependence (rather than participation, discussed in the last section).[8] Various notions of ontological dependence are widely discussed in contemporary philosophy and it is arguable that we now understand them better. In particular, Fine's notion of *essential* dependence looks especially helpful where platonic universals are concerned. Fine developed this notion in response to the inadequacies of purely modal conceptions of ontological dependence. To say that one thing (call it x) depends *in a purely modal sense upon* another (y) is to say that necessarily, if x then y. Fine (1995) argues, however, that there are many cases in which the purely modal construal of dependence is insufficient for capturing the ontological dependence involved. For instance, strictly speaking Socrates is modally dependent on his singleton set, but surely Socrates is not ontologically dependent on his singleton set; rather, the set depends for its existence on Socrates and not vice versa. In such cases, Fine holds that we need an alternative notion to capture this sort of ontological dependence, namely, *essential dependence*. At the heart of the notion of essential dependence is the idea that, in many (if not all) cases of ontological dependence, 'the necessity of the conditional 'x exists only if y does' should be appropriately tied to the nature of the dependent entity item x' (Fine 1995, 272–273).[9]

---

[8] It is well worth noting at this point that relations of ontological dependence should not be thought of as universals. To use Lowean terminology, ontological dependence is a formal rather than a material phenomenon. By this it is meant that ontological dependence relations are not items *within* an ontological category but rather capture the dependences *between* the ontological categories (in the platonic case, the categories of abstract universals and concrete states of affairs). Importantly, because ontological dependence is not itself a universal, it is not subject to the sort of regress problem discussed earlier (section 3) in connection with relational realism.

[9] It is not my intention here to enter into a detailed discussion of the Finean notion of essential dependence. An explanation of the basic idea suffices for current purposes. For further details, interested readers should see Correia 2005, Fine 1995 and Lowe 1998, Ch. 6. Note that although Lowe's notion of ontological dependence is similar to Fine's notion of essential dependence (Lowe 1998, 149), Lowe prefers to cash it out using the notion of identity rather than essence. Again, the argument to follow is unaffected by these specific details.

The reason why platonism would be best cashed out using the notion of essential dependence rather than purely modal dependence is this. Since platonic universals are transcendent entities, existing outside of space and time, they plausibly exist necessarily (see e.g. Bird 2007, Ch. 5 and Tugby 2013). But as critics of the modal account of ontological dependence point out, where the dependee is a necessary existent, anything whatsoever will be dependent on it in the modal sense.[10] Clearly, this leaves the platonist with too many ontological dependences. If ontological dependence is understood in a purely modal sense, it means that, for instance, the existence of an electron particle is dependent on the proton universal, and the universal of being ice-cream, and whatever other transcendent universals there are. In order to avoid these implausible claims, what a plausible form of platonism should say is that the ontological dependence is appropriately constrained by the *nature* of the entities involved, so that, for example, the existence of an electron is dependent on the electronhood universal and no other. The platonic thesis would then be drawn out along the following Finean lines:

Platonic dependence: Essentially, for each particular, if it has some property P, then there is a transcendent universal P.

How does all of this bear upon the one over many problem? Well, importantly, it seems that essential dependences are especially well suited to ground *explanations*. For instance, to return to the Socrates example above, it seems the existence of Socrates plausibly explains the existence of Socrates' singleton set (rather than vice versa), and what seems to ground this explanation ontologically is the fact that Socrates' singleton set essentially depends on Socrates (rather than vice versa). To return to a point made earlier, it is generally the case that a depender is explained by the dependee but not vice versa. If we are to provide a metaphysical explanation for one over many facts – facts of the form a is F – then platonic dependence relations seem to run in precisely the appropriate direction. This is because, in line with the platonic dependence principle above, instantiation facts essentially depend on the relevant universals rather than vice versa.

How, then, can it be that a is F while b is also F? Roughly, a platonic analysis of one over many could be as follows: the fact of a's being F is essentially dependent on universal F, while b's being F is also essentially dependent on universal F. Thus, sameness of type arises when there is sameness of essential dependence

---

10 To use an example from Correia (2005, 30), assuming that Socrates is a contingent existent and that the empty set is necessary, then the former is modally dependent upon the latter. And yet, surely we do not want to say that Socrates depends for his existence on the empty set.

to a given universal. This is made possible by the fact that relations of essential dependence can be many-one, which is to say that distinct states of affairs can be essentially dependent on the same universal. So, to generalise, we may say that one over many phenomena are explained by the fact that platonic relations of essential dependence are *many-one*.

I have only been able to provide a brief sketch here of how a Platonist ought to set about answering the one over many problem (for further details about the platonic solution to the problem, see my forthcoming article 'Universals, Laws, and Governance'). But if the main thrust of the argument is correct, it shows that it is not surprising that Plato, who was arguably the first to identify the one over many problem, rejected the immanent view of universals in favour of the transcendent view.

To sum up, if the arguments of this chapter are correct, then Armstrongian immanent realists are indeed mirage realists, at least to some extent. In contrast, it is far from clear that platonic realists who take the one over many problem seriously need be guilty of this charge.

Bibliography

Armstrong, D. M. 1978. *Universals and Scientific Realism, Vol. 1, Nominalism and Realism*. Cambridge: Cambridge University Press.
Armstrong, D. M. 1980. "Against 'Ostrich Nominalism': A Reply to Michael Devitt". *Pacific Philosophical Quarterly*, 61, 440–9.
Armstrong, D. M. 1989. *A Combinatorial Theory of Possibility*. Cambridge: Cambridge University Press.
Armstrong, D. M. 1991. "Classes Are States of Affairs". *Mind*, 100, 189–200.
Armstrong, D. M. 1997. *A World of States of Affairs*. Cambridge: Cambridge University Press.
Armstrong, D. M. 2004. *Truth and Truthmakers*. Cambridge: Cambridge University Press.
Armstrong, D. M. 2005a. "How Do Particulars Stand to Universals?" In Zimmerman, D. (ed.), *Oxford Studies in Metaphysics, Vol. 1*. Oxford: Oxford University Press, 139–54.
Armstrong, D. M. 2005b. "Four Disputes about Properties". *Synthese*, 144, 309–20.
Armstrong, D. M. 2005c. "Reply to Simons and Mumford". *Australasian Journal of Philosophy*, 83, 271–6.
Armstrong, D. M. 2006. "Particulars Have Their Properties of Necessity". In Chakrabarti, A. and Strawson, P. F. (eds.), *Universals, Concepts and Qualities: New Essays on the Meaning of Predicates*. Burlington: Ashgate, 239–48.
Baxter, D. 2001. "Instantiation as Partial Identity". *Australasian Journal of Philosophy*, 79, 449–64.
Bird, A. 2007. *Nature's Metaphysics: Properties and Laws*. Oxford: Oxford University Press.
Bynoe, W. 2011. "Against the Compositional View of Facts". *Australasian Journal of Philosophy*, 89, 91–100.
Campbell, K. 1990. *Abstract Particulars*. Oxford: Basil Blackwell.

Correia, F. 2005. *Existential Dependence and Cognate Notions*. Munich: Philosophia.
Correia, F. 2008. "Ontological Dependence". *Philosophy Compass*, 3, 1013–32.
Devitt, M. 1980. "'Ostrich Nominalism' or 'Mirage Realism'?" *Pacific Philosophical Quarterly*, 61, 433–9.
Devitt, M. 2010. "Postscript to 'Ostrich Nominalism' or 'Mirage Realism'?" In *Putting Metaphysics First: Essays on Metaphysics and Epistemology*. Oxford: Oxford University Press.
Dodd, J. 1999. "Farewell to States of Affairs". *Australasian Journal of Philosophy*, 77, 146–60.
Fine, K. 1982. "Dependent Objects". Unpublished manuscript.
Fine, K. 1995. "Ontological Dependence". *Proceedings of the Aristotelian Society*, 95, 269–90.
Lewis, D. 1983. "New Work for a Theory of Universals". *Australasian Journal of Philosophy*, 61, 343–77.
Lowe, E. J. 1998. *The Possibility of Metaphysics*. Oxford: Oxford University Press.
Mumford, S. 2007. *David Armstrong*. Stocksfield: Acumen.
Oliver, A. 1996. "The Metaphysics of Properties". *Mind*, 105, 1–80.
Plato 1975. *Phaedo*. Gallop, D. (Trans.), Oxford: Oxford University Press.
Rissler, J. 2006. "Does Armstrong Need States of Affairs?". *Australasian Journal of Philosophy*, 84, 193–209.
Tugby, M. 2013. "Nomic Necessity for Platonists". *Thought*, 2, 324–31.
Tugby, M. Forthcoming. "Universals, Laws, and Governance". *Philosophical Studies* (DOI: 10.1007/s11098-015-0521-2).
Vallicella, W. 2000. "Three Conceptions of States of Affairs". *Noûs*, 34, 237–59.

Francesco F. Calemi
Ostrich Nominalism or Ostrich Platonism?

1 What is it like to be an ostrich?

"Ostrich" is an epithet that Armstrong employs to label what he considers the weakest form of contemporary Nominalism insofar as it refuses to engage in serious ontological enquiry. This epithet – and the negative judgment that comes with it – has deeply influenced the debate between nominalists and realists, and it is supported by three major arguments put forward by Armstrong, each of which bears on certain drawbacks that allegedly cripple Ostrich Nominalism: the *argument from gross facts*, the *harlot argument*, and the *truthmaker argument*. Deploying this array of arguments, Armstrong aims to show that, in confronting with the problem of the existence of properties, Ostrich Nominalists must do something else than just burying their heads in the sand. In what follows, I will first review Armstrong's three arguments, contending that none of them is fully satisfactory. Then I will sketch a Platonic theory of predication, and I will argue that such a theory, while sharing the "ostrich" feature of Ostrich Nominalism, provides a more adequate response to Armstrong's three challenges.

2 The argument from gross facts

Let us start our discussion by scrutinizing the argument from gross fact. In his masterpiece *Universals and Scientific Realism* Armstrong introduces this argument by asking the following allegedly compelling "One over Many" question: *how can two or more things be the same?* Let us consider Armstrong's example:

> How can two different things both be white or both be on a table? (Armstrong 1978a, 12)

At first sight this question sounds odd for such trivial facts don't seem to demand any explanation: one could ask what's so strange about two numerically different things being both white, or being both on the same table? To understand why Armstrong believes that the One over Many problem is a *genuine* problem we should first pay attention to his conception of the nature of the debate between realists and nominalists. Armstrong holds that the ontological disagreement between these two parties is based upon

---
Francesco F. Calemi: University of Perugia, email: francesco.calemi@unipg.it

> a basic agreement [...]: that in some minimal or pre-analytic sense there are things having certain properties and standing in certain relations. (Armstrong 1978a, 11)

More precisely, despite their remarkable divergent views, both the realist and the nominalist accept what seems to be a triviality:

> The piece of paper before me is a particular. It is white, so it has a property. It rests upon a table, so it is related to another particular. Such gross facts are not, or should not be, in dispute between Nominalists and Realists. (Armstrong 1978a, 11)

The agreement underlying the disagreement among realists and nominalists is about this "gross fact"[1]: if this paper is white, then it has the property of being white, so it has a property.[2] This contention is justified by the fact that, according to Armstrong, "the sentence-type 'Pa' expresses the [...] proposition that a has the property P." In more specific terms, Armstrong holds that the sentence "Pa" is true if and only if its sub-sentential expressions "P" and "a" correspond, respectively, to a property, P, and a particular, a, and a instantiates P.[3] It follows that an elementary predication of the form

(Elementary-P) a is P

has to be analyzed in terms of property instantiation as follows

(Property-I) a instantiates the property P (or P-ness).[4]

Armstrong takes for granted that this analysis cannot be proved; rather, the tenet that for a particular to be in a certain way is for it to instantiate a correlative property should be taken as a pre-analytical *datum* that everybody must accept beyond the reach of reasonable controversy. As such, Armstrong calls it a *gross fact*. Once this has been accepted, gross facts thrive everywhere: "Socrates is wise" is true in so far as Socrates instantiates wisdom; "e is an electron" is true in so far as e instantiates electronhood; "a has mass M" is true in so far as a instantiates having mass M, and so on. Since everyone must accept as a *datum* the truth-conditional equivalence between (Elementary-P) and (Property-I) regardless of her preferred

---

[1] Armstrong 1978a, 11.
[2] The same goes for relations: if the paper is on the table, then the relation of *being on* holds of the ordered pair constituted by the paper and the table; hence there is a relation between the paper and the table. Even if this work shall focus only on properties, our remarks will be valid *mutatis mutandis* also for relations.
[3] This is clearly stated in Armstrong 1978b, 21.
[4] This is has been highlighted also by Oliver 1996, and Rodríguez-Pereyra 2000.

ontological credo, (at least some of) the outcomes of the application of the following analysis-schema constitute gross facts that "are not in dispute":[5]

(Sch) ⌜*a* is *P*⌝ is true iff *a* instantiates *P*-ness.

Armstrong's first argument against Ostrich Nominalism proceeds precisely from this premise: if the debate between Nominalists and Realists is to be considered a serious debate, (at least some of) the substitution instances of the schema (Sch) must be accepted by the competing factions. Moreover – Armstrong continues – the Nominalist and the Realist

> can agree that the paper is white and rests upon a table. It is an adequacy-condition of their analyses that such statements come out true. (Armstrong 1978a, 11)

But if they accept the truth of an elementary predicative statement having a one-place predicate, such as

(1)   *a* is white,

then they also have to accept – as a gross fact – its ontologically loaded counterpart, i.e. the statement enfolding a two-place predicate and an explicit property-designator

(2)   *a* instantiates whiteness.

Let us concede all this for the sake of the argument, and ask once again: where does the One over Many problem come from? What does it mean to ask how two different things can both be white? Armstrong's answer echoes Plato's astonishment toward the "amazing statement" that "the many are one and the one many" (*Philebus*, 14C):

> as Plato was the first to point out, this situation is a profoundly puzzling one, at least for philosophers. The same property can belong to different things [...] Apparently, there can be something identical in things which are not identical. Things are one at the same time as they are many. How is this possible? (Armstrong 1978a, 11)

If one assumes that it is a pre-analytical gross fact that elementary predication is to be construed as property instantiation, and if two different non-overlapping things are white, then they both instantiate the same whiteness; and if they instantiate the same whiteness, then there is something identical about them, i.e.

---

[5] Armstrong 1978a, 11.

they share the same property. So the question is how is this "sharing" possible, given their numerically distinctedness and their being spatially non-overlapping?

In particular, one may ask whether and how the nominalist may accept (Sch) while at the same time avoiding its ontological implications. Armstrong reviews seven attempts to neutralize the ontological weight of gross facts about predication: Predicate Nominalism, Concept Nominalism, Mereological Nominalism, Class Nominalism, Natural Class Nominalism, Resemblance Nominalism, and Ostrich Nominalism. Yet he takes the first six of these as serious ontological strategies, whereas the last one is presented as a "Cloak-and-Dagger" theory. While it might be of interest to go into details of each nominalist position, lack of space prevents us from doing so here. Rather I will simply present their main features that will prove relevant to our concerns.

According to Armstrong, serious Nominalists can dialectically react to the One over Many problem by providing *reductive analyses*. Most Nominalists reject the idea that property instantiation is primitive, and try to *further analyze* it in order to reduce property instantiation to notions that do not imply the puzzling existence of sharable properties. Consider, by way of example, Predicate Nominalism. Whoever espouses this strategy acknowledges, on the one hand, that if a particular – say a – is in a certain way – say F – , then it possesses a corresponding property, i.e. the property of being F; but, on the other hand, she also holds that to have a property is nothing but to satisfy a certain predicate, i.e. the predicate "(is) F". And since it is impossible to deduce the existence of genuine properties from the statement "a satisfies the predicate '(is) F'", the Predicate Nominalist concludes that a true statement of the form "a is F" does not, deep down, commit us to genuine properties: "property talk" is nothing but "predicate talk". *Mutatis mutandis* the same holds true for every other form of Nominalism, with the exception of the Ostrich Nominalism.

Unlike the other kinds of Nominalism, Ostrich Nominalism – Armstrong claims – is not a serious position, for its advocates, such as Quine,

> refuse to countenance universals but [...] at the same time see no need for any reductive analyses of the sorts just outlined. There are no universals but the proposition that a is F is perfectly all right as it is. (Armstrong 1978a, 16)

As I take it, here Armstrong purports to raise a dilemma that the Ostrich Nominalist does not intend to resolve. Before discussing the dilemma it should be noted that, even if in the above passage Armstrong wrongly depicts the Ostrich Nominalist as a *universals denier*, his purpose is to criticize the Ostrich Nominalist as a *properties denier*. Thus Armstrong's statement should be charitably interpreted as expressing the claim that, according to the Ostrich Nominalist, there are no properties but the proposition that *a* is *F* is perfectly all right as it is. Now let us assume that the statement "Socrates is wise" is a predicative statement that anyone would accept as true, whether Realist or Nominalist, even Ostrich Nominalist. Under this hypothesis, the Ostrich Nominalist would hold both of the following:

(i) There are no properties.

(ii) "Socrates is wise" is true.

According to Armstrong, however, (i) and (ii) are jointly inconsistent. If "Socrates is wise" is true, then its truth implies the truth of "Socrates instantiates wisdom" (by (Sch)); and if this latter statement is true, then there is something that Socrates instantiates, i.e. wisdom. Now, wisdom is, if anything, a property; but this is inconsistent with assumption (i), which denies the existence of properties altogether. In turn, if wisdom does not exist, it is as impossible for "Socrates is wise" to be true as it is for "Socrates instantiates wisdom". Indeed, the latter's falsehood implies the falsehood of "Socrates is wise" altogether. But this is inconsistent with the assumption (ii). So either (ii) is false, contrary to our innocuous assumption, or else it is (i) that must be false. More generally, either no elementary predicative statement could ever be true, or else (i) must be dropped.

However, the Ostrich Nominalist is not as "ostrichy" as painted by Armstrong. Indeed Armstrong is only able to raise the charge of inconsistency against Ostrich Nominalism by overlooking some of its features. To illustrate the point let us return to the true statement

(3) Socrates is wise.

In virtue of (Sch), we can infer from (3) the statement

(4) Socrates instantiates wisdom.

So far, so good. But, as Armstrong himself claims, the Ostrich Nominalist "refuse[s] to countenance universals." If this is right, then for the Ostrich Nominalist the statement (4) is not true but *simply false*; yet the Ostrich Nominalist is not thereby obliged to assert the falsity of (3) as well. After all – the Ostrich Nominalist may reply – "gross facts" are just *prima facie* evidence, and as such they can be emended or, if necessary, denied. For an Ostrich Nominalist, it is more advisable to reject (Sch) because, from her point of view, it is impossible for a true elementary predicative statement to be truth-conditionally equivalent to a property-instantiation statement: while there are true elementary predicative statements, their loaded counterparts are always false (since their truth requires the existence of entities that don't exist, i.e. properties). This is why, in denying the truth of (4), the Ostrich does not mean to deny altogether the truth of (3). The following well-known passage by Quine illustrates this kind of nominalism at work:

> One may admit that there are red houses, roses, and sunsets, but deny, except as a popular and misleading manner of speaking, that they have anything in common. The words "houses", "roses", and "sunsets" are true of sundry individual entities which are houses and roses and sunsets, and the word "red" or "red object" is true of each of sundry individual entities which are red houses, red roses, red sunsets; but *there is not*, in addition, any entity whatever, *individual or otherwise*, which is named by the word "redness", nor, for that matter, by the word "household", "rosehood", "sunsethood". (Quine 1948, 10. Italic mine)

If this is correct, then Armstrong does not pay due attention to the Nominalism inspired by Quine: at most the Ostrich Nominalist, rather than a dilemma, must face a *trilemma* that is far from puzzling:

(i) There are no properties.

(ii) "Socrates is wise" is true.

(iii) "Socrates is wise" is true iff Socrates instantiates wisdom.

Rejecting (iii) while retaining (i) and (ii) is a perfectly legitimate and consistent move. Although Armstrong presupposes that statements of the form "*a* is *P*" cannot be true unless one admits an ontology of properties, the Ostrich Nominalist shows that it is possible to give an adequate account of the truth of elementary predicative statements without such an ontological commitment. From the Ostrich point of view, the One over Many question, "How can two different white things, *a* and *b*, both be white?", eventually turns out to be trivial, and as such it deserves just a trivial answer: *a* and *b* are both white *because a* is white and *b* is white. That is the way it is, and from a metaphysical point of view there is nothing left to be explained.

3 The harlot argument

So we come to Armstrong's second argument against Ostrich Nominalism: the harlot argument. As Armstrong contends, the Ostrich Nominalist gives to predicates

> what has been said to be the *privilege of the harlot*: power without responsibility. The predicate is informative, it makes a vital contribution to telling us what is the case [...] yet ontologically it is supposed not to commit us. Nice work: if you can get it. (Armstrong 1978a, 16. Italic mine)

As we can see, in this passage the refusal of Ostrich Nominalism has no bearing on the acceptance of the gross truth expressed by (Sch), but rather on the fact that, according to Quine, predicates – their descriptive "power" notwithstanding – simply do not convey any ontological commitment. It is worth recalling that this Quinean thesis about predicates is not the product of an arbitrary idiosyncrasy, but the consequence of Quine's theory of quantification. According to Quine "to be is to be the value of a [bound] variable".[6] The central idea expressed by this motto is that for something to exist is for it to belong to the domain quantified over by objectually construed first-order quantifiers. Roughly, in a regimented language there is ontological commitment wherever there is quantification, and there is quantification wherever there is ontological commitment. But there is quantification only if it is possible to introduce bound variables into the places occupied by referential terms, and Quine contends that predicates are not referential terms.[7] If predicates were genuine referential expressions, they might be replaced by coreferential terms *salva veritate* and *salva significatione*. Let us suppose that, given statement (3), the relevant predicate "(is) wise" is actually a linguistic device that names the same property being named by the abstract singular term "wisdom", that is the property of being wise. If this were true, then such a predicate and the name "wisdom" would be coreferential; and if they were coreferential terms, then, given (3), the former might be replaced by the latter *salva veritate* and *salva significatione*. But this is false as the sequence of words "Socrates wisdom" is utterly meaningless. Quine believes that this shows that predicates are non-referential expressions; as such they do not support quantification; thus they do not carry any ontological commitment, and *a fortiori* their use does not commit us to sharable properties. The following theses summarise the main premises of Quine's argument:

(*M*) To be is to be the value of a bound variable.

(*LG*) It is impossible to introduce bounded variables into predicative positions.

In both volumes of *Universals and Scientific Realism* the real meaning of Armstrong's criticism of Quine's thesis that predicates lack of ontological seriousness is far from clear. At first sight, it appears that Armstrong's animadversion about Quine's stance has far more to do with (*LG*). At least, the question seems to be precisely this in *Universals and Scientific Realism*, where Armstrong explicitly demands the introduction of second-order quantification. In a brief passage Armstrong acknowledges that from the schematic statement

---

[6] Quine 1948, 15.
[7] See Quine 1970, and also Searle 1969 for more details.

(Elementary-P*) Fa

one can validly infer not only the following first-order quantified statement

(5) $\exists x(Fx)$

but also a second-order quantified statement as

(6) $\exists X(Xa)$

As an upholder of the non-existence of uninstantiated universals, Armstrong complains that the standard symbolism is "potentially misleading":[8]

> The symbolism of the first inference suggests the doctrine of the particular without its properties, that of the second the doctrine of uninstantiated properties. (Armstrong 1978a, 110)

Accordingly he proposes an emendation of the symbolism that consists in replacing the variable with an en-dash between brackets, "(–)", so as to obtain such open formulae as: "$F(–)$" and "$(–)a$". The revised notation is supposed to be closer to the demands of Armstrong's Realism because "it makes clear that what we are dealing with the whole time is a particular-having-certain-properties".[9] But beyond this technical gloss, Armstrong does not provide us with an answer to Quine's argument to the effect that second-order logic is not possible, thus remaining silent about an issue that should instead play a crucial role into his attempt to dispose once and for all with Ostrich Nominalism. Neither does Armstrong hold that predicates are genuine referential terms; and although he continuously talks about a certain "correlation" holding between predicates and properties, such a tie cannot be construed as a one-one referential relation – not without declaring the falsity of the *sparseness thesis*, central to Armstrong's metaphysics:

SPARSENESS THESIS: It is not the case that every predicate applies to the particulars it applies to in virtue of some property. (Armstrong 1978a, 39)

A few years after the publication of *Universals and Scientific Realism*, Armstrong reasserted once again that what prevents Quine from appreciating (and thus to feeling the urge to solve) the One over Many problem is

---

[8] Armstrong 1978a, 110.
[9] *Ibidem*.

his extraordinary doctrine that predicates involve no ontological commitment. In a statement of the form "F*a*", he holds, the predicate "F" need not to be taken with ontological seriousness. (Armstrong 1980, 105)

But, once again, Armstrong's text is almost entirely lacking in clear arguments for the legitimacy of second-order quantification, or the conclusion that predicates occupy syntactic places accessible to quantification. After all there is a clear reason why this cannot be a viable solution for Armstrong, at least not without the obligation to abdicate to his sober sparse Realism: if second-order quantification is allowed, then *every* predicative expression can occupy positions accessible to quantification; and if this were true, clearly *every* predicate that truly applies to something picks out a correlative property, no matter its syntactical complexity. In the end, even Armstrong himself cannot afford to take predicates with the same ontological seriousness that he nonetheless demands from the Ostrich Nominalist.

4 The truthmaker argument

Over the past decade there has been a truthmaking turn in ontology, and Armstrong has unquestionably been fully engaged in it. As a truthmaker theorist Armstrong holds that there is a substantial (i.e., non trivial) relationship holding between *what there is* and *what is true*, this relationship being expressed by the so-called *truthmaker principle*: for every true proposition, p, there must be a truthmaker, viz. something in the world that is such that it makes it true that p:

(TM-Principle) $\forall p \Box (p$ is true $\leftrightarrow \exists x(x$ makes p true$))$

Note that the truthmaking job is not to be understood as a form of causality, but in terms of *grounding*. For something, x, to be the truthmaker of a true proposition, p, is for it to be something *in virtue of which* p is true. As such, truthmaker theory provides ontological explanations of truths' truth: an ontological explanation of a true proposition is reached in pointing out what actual entity grounds its truth. Truthmaking also requires a non-propositional necessitation holding between a truth and its truthmaker, so that

(TM-Necessitarianism) x makes p true → $\Box(x$ exists → p is true).

Truthmaking is non-propositional, since if p is made true by T, then the truthmaking relation does not hold between two propositions, p and <T exists>, but between a proposition, p, and a non-propositional entity, T.

Armstrong's by now classical truthmaker theory is a formidable challenge to Quine's criterion of ontological commitment:

> Accepting the truth-maker principle will lead one to reject Quine's view [...] that predicates do not have to be taken seriously in considering the ontological implications of statements one takes to be true. (Armstrong 1989, 89)

> Why should we desert Quine's procedure for some other method? The great advantage, as I see it, of the search for truthmakers is that it focuses us not merely on the metaphysical implications of the subject terms of propositions but also on their *predicates*. (Armstrong 2004, 23)

On the face of it, Armstrong offers his truthmaker theory as a new method for ontological inquiry, and a means to keep it *serious* and *honest*.[10] His truthmaker argument against Ostrich Nominalism shows how such a theory is supposed to work. The exposition of this argument is in order.

We have assumed that the Ostrich Nominalist takes (3) to be true. But if (3) is true, then there must be something in the world in virtue of which it is true (by the TM-Principle). Armstrong's argument then continues by raising the usual truthmaker question: what makes it true that Socrates is wise? Since the Ostrich Nominalist countenances only particulars, he will typically reply that the particular labelled as "Socrates" is (3)'s truthmaker. But, as Armstrong teach us, particulars can be thought of as either *bare* or *thick*. A *bare particular* is a "blob", as Armstrong likes to say, that is a particular conceived as absolutely lacking of any properties, while a *thick particular* is a particular that is thought of as "already possessing its properties".[11] So in invoking Socrates as the truthmaker for (3), either Ostrich Nominalists construe it as a thin particular, or as a thick particular. On the one hand, if the entity denoted by "Socrates" is a thin particular, then the Ostrich Nominalists are providing us with an *insufficient truthmaker* because Socrates' existence is compatible with both (3)'s being true, and (3)'s being false (i.e., Socrates exists in worlds where he is wise, and in worlds where he is not wise); this patently constitutes a violation of TM-Necessitarianism. On the other hand, if Socrates is construed as a thick particular, then the Ostrich Nominalists are supplying a *redundant truthmaker*, that is a truthmaker that while necessitating (3)'s truth fails to satisfy a *minimality constraint* according to which whatever makes (3) true should include only properties that are relevant.[12] So, in taking Socrates as a candidate for playing the truthmaking job required by (3)'s truth, ei-

---

10 Cf. Armstrong 2004, 43.
11 Armstrong 1978a, 114.
12 Cf. Armstrong 2004, 19–21.

ther the Ostrich Nominalist furnishes an insufficient truthmaker, or a redundant one.

The argument can be pushed further by a proof by cases: whoever provides insufficient truthmakers cheats in so far as they take as ontological ground of a true proposition something that does not ground it at all.[13] Moreover, if the Ostrich Nominalist suggests that Socrates *qua* thick particular is (3)'s truthmaker, then – once again – she is cheating, as she would covertly countenance something she cannot, i.e. properties. This is why the Ostrich Nominalist cheats anyway, that is, she does not seriously address the ontological problem at issue.

The truthmaker argument outwardly rises an outstanding challenge to the Ostrich Nominalism. Yet two remarks are worth making. For one thing, it is doubtful whether the Ostrich Nominalist can go along with one of the premises on which the argument rests, namely, that the entity denoted by a name like "Socrates" must be either a thin particular or a thick particular. That premise is unacceptable if it contains a false dilemma; and it contains a false dilemma if the distinction between thin and thick particulars somehow *presupposes* the existence of properties. But in Armstrong's terminology a thick particular is "a thing taken along with all its properties",[14] while a thin particular is "a thing taken in abstraction from all its properties":[15] in both cases properties are plainly "alive and kicking" – so to say. So, unless the Ostrich Nominalist's ontological credo is negotiable, she can persist in holding that properties do not exist. And if properties do not exist, they can neither be "taken along with", nor "abstracted from", a particular thing. Thus, from the Ostrich point of view, the thin/thick distinction should be rejected: the premise invites the Ostrich Nominalist to start a game that she simply cannot play.

Anyway, even if some mild solutions can be reached by nominalists,[16] it is hard to imagine how Ostrich Nominalists may react to the truthmaker argument without rejecting outright the TM-Principle. Does this remain a viable option for them? There is a sense in which the upholder of the TM-Principle asks *too much*. This is not because there cannot be a relationship between elementary predicative truths and reality for an Ostrich Nominalist. Of course there is: it's just that

---

[13] This does not mean that Socrates, as a thin particular, is not relevant to the truth of (3), but that there is more to (3)'s truthmaker than just a thin particular. Armstrong thinks that such a truthmaker is a state of affairs, *Socrates' instantiating wisdom*, that enfolds both a thin particular and a property as its constituents, but that is not analyzable without remainder into them.

[14] Armstrong 1978a, 114.

[15] Armstrong 1978a, 114.

[16] There are, for instance, nominalists who reject TM-Necessitarianism, such as Melia 2005, while others uphold a counterpart theory for particulars, as Rodríguez-Pereyra 2002.

is not to be framed in truthmaking fashion. In truthmaker theory *only existence matters*; but according to the Ostrich Nominalist, it is not the case that such predicative truths as (3) are true *solely by virtue of the existence of some entity*. As we have seen, the Ostrich Nominalist contends that (3) is true if and only if (i) Socrates exists and (ii) he is wise: while condition (i) is concerned with *what there is* – call this the *existential condition* –, condition (ii) regards *how it is* – call this the *descriptive condition*. Satisfying these conditions is certainly enough to give a quite acceptable and modest explanation for (3)'s truth:

"Socrates is wise" is true because Socrates is wise,

but it is not sufficient if one assumes that to explain a truth we must appeal *only to what there is* – as the Armstrongian truthmaker theory imposes. Therefore, for an Ostrich to accept the TM-Principle would mean for her to be forced to *reify* descriptive conditions: but to reify descriptive conditions is, or implies, either to nominalize predicates, or to admit second-order quantifiers, and we have seen before that the Ostrich Nominalist rejects both these moves.

5 Sketch for a Platonic theory of predication

I have offered some answers on behalf of the Ostrich Nominalist to the Armstrong's arguments. Yet there are still some problems for Ostrich Nominalism: one of these is represented by the phenomenon of *predicate anaphora*. In discussing this issue, I will briefly sketch a Platonic theory of predication that meets the predicate anaphora challenge, and I will consider it in the light of the three main issues raised by Armstrong's arguments previously exposed showing that the version of Platonism I prefer employs the "ostrich strategy".

For one thing, the Platonism I prefer shares with Ostrich Nominalism the contention that we do not have to posit properties as ontological correlates of predicates either to explain why an elementary predication is true, or to account for the multiple applicability of predicates. (More on this below.) Nevertheless, as I have argued elsewhere (Calemi 2012; 2014), properties should be acknowledged as ontological correlates of predicates in order to explain why and how the phenomenon of predicate anaphora is possible.[17] Take the complex sentence

(7) (a) John is honest and
(b) this is a property that every good politician should have.

---

[17] On the predicate-anaphora see also Swoyer 1999, and Künne 2006.

On the face of it, here the pronoun "this" in (b) is anaphorically tied to the adjective "honest" in (a). More precisely, "this" refers back to something that is *introduced* (even if not *mentioned*, nor *named*) by "honest", namely, the property of being honest, or honesty. Indeed, given the context provided by (a), the sentence in (b) can be restated as

(7) (b′) Honesty is a property that every good politician should have.

Since in (b) "this" is anaphoric on "honest", the reference of the former must be the same as is the reference of the latter. But the pronoun "this" occurs in a *higher-level predication*, so it plays the same referential role as an abstract singular term, i.e. an *explicit property-designator*. Therefore, the reference of "this" is a property; hence the reference of "honest" is a property too, namely honesty. This does not contradict the fact that general terms are *predicable*. It barely means that general terms (such as "honest") have two different semantic roles: besides being referentially tied to the things they are *true of*, general terms *stand for* (or *connote*) properties.[18] Since this holds for any genuine general term, I endorse the thesis that any genuine general term picks out a matching property, no matter how complex it is. So, *contra* Quine, even if general terms are not names of properties, and even if syntax prevents us from treating them as names (in that, given a predicative sentence, we cannot replace a general term and its correlative nominalization *salva significatione*), they are nonetheless *bona fide* property-referring expressions.

Of course, in natural language the mere juxtaposition of a singular term and a general term does not yield a well-formed elementary predication: terms must be copulated, and copulation is the proper function of the copula. Being a *predicate-forming operator* on general terms, the copula outputs predicates; in turn, predicates behave as a sentence-forming operators on singular terms, and inherit their double semantic function from the general terms they are composed of: predicates are *true of* the things that belong to their extensions, and *connote* a property just as general terms do. But even if any genuine general term connotes a correlative property, all this is entirely irrelevant when it comes to explaining the truth of a subject-predicate sentence, for the only reference that matters in this case is that made by the corresponding subject term:

⌜*a* is *P*⌝ is true iff *a* is *P*.

In as much as ordinary predicates play two different semantic rôles, their complex semantics has to be somehow brought into a perspicuous schema that aims to

---

18 Several authors have taken the view that general terms *connote*, *specify*, *express*, or *introduce* properties: Mill 1843, Wolterstorff 1970, Loux 1978, Wiggins 1984, Strawson 1987, Künne 2006.

represent their form. This lead us to exclude that predicates construed as *purely extensional expressions* are fit to serve our notational needs: the referential role of an ordinary predicate does not merely boil down to its *being true of* something. Nor may ordinary predicates be represented by such explicit property-designators as "*F*-ness", "*G*-ness" and so on: indeed, *explicit property-designators* have the same logical behavior as individual constants, and individual constants are not predicable at all, unlike ordinary predicates. As far as predicates actually introduce properties into the discourse domain neither by naming them, nor by mentioning them, and have *predicability* as an irreducible feature, predicates should be rather construed as *property-connoting expressions*. Let us assume the following notation to mark the difference between the three types of linguistic items indicated above:

Purely extensional predicates	$F, G, H \ldots$
Property-connoting predicates	$F_\phi, G_\phi, H_\phi \ldots$
Explicit property-designators	$\phi_F, \phi_G, \phi_H \ldots$

Given these notational conventions, a Platonic schema that perspicuously represents ordinary predication should be rendered in a language, say $\mathcal{L}_i$, having as primitive predicative terms property-connoting predicates, rather than purely extensional predicates:

(Elementary-P#) $P_\phi a$

Let us develop the point a little further by asking how we are supposed to quantify over predicates' *connotata*. The debate offers two main alternatives: the commitment that predicates harbor could be made explicit either by way of regimented second-order quantification, or by way of regimented first-order quantification.[19] The former solution is the one that the Platonism I prefer endorses: $\mathcal{L}_i$ is a second-order language. This leaves no option but to address Quine's case against second-order quantification. And as far as I can see, Quine's argument assumes that we introduce something into the discourse domain only by *naming it*; therefore if predicates are genuine referring expressions, then they name something (hence they could be substituted with coreferential names). But against the background that I have sketched thus far, this presupposition proves to be false: predicates *connote* their referents, rather than naming them, and they are just what they are, i.e. predicates, not names.

---

19 A typical example of the first type can be found in Cocchiarella 1986, while a typical example of the second type can be found in Bealer 1982.

Anyway, whether one should adopt second-order logic is a controversial issue, and there could be Platonists who prefer to avoid doing so. These Platonists should hold on the one hand that predicates carry an ontological commitment to properties, and on the other hand that ontological questions must be framed only in the apparatus of first-order logic. And since first-order logic is blind to the predicates' connotata, their solution is to bring into play a schema that renders in an extensional language such an ontological commitment. The schema (Sch) yields statements with explicit property-designators from elementary predications that are true in a language into which predicates are construed as property-connoting expressions: (Sch) is, so to say, a bridge schema that *turns connotata into nominata*. (The schema (Sch) should also convey additional information concerning both its object language, and its metalanguage, as we will see in a while.) Moreover, it should be emphasized that rendering ordinary predications' commitment in a first-order language, say $\mathcal{L}_e$, brings about an impoverishment of their referential role, along with a loss of their predicability: for a first-order language item to introduce something into the discourse domain is for it to occupy a position accessible to first-order quantifiers, and in first-order logic one can only quantify in the positions occupied by individual constants; and individual constants cannot play any predicative role. Anyway in $\mathcal{L}_e$ predicability is reintroduced by the instantiation predicate constant "*Ins*" which plays a unifying function among the sub-sentential elements of the property-instantiation sentences. So, assuming $\mathcal{L}_e$ as a metalanguage, the following schema will hold:

(Sch⁻) $\ulcorner P_\phi a \urcorner$ is true in $\mathcal{L}_i$ iff $Ins(a, \phi_P)$.

In other words, in order to render the commitment that property-connoting predicates bear in $\mathcal{L}_i$ (Sch⁻) splits up their complex semantics into two different roles played by two, rather than one, linguistic items: the reference that in $\mathcal{L}_i$ is introduced by the property-connoting predicate "P_ϕ", in $\mathcal{L}_e$ is introduced by an explicit property-designator, while "P_ϕ"'s predicative function (i.e. its *being true of*) is delivered in $\mathcal{L}_e$ by the predicate constant "*Ins*" that links into propositional unity two (or more) individual constants. (Note that in this way "$P_\phi x$" and "$Ins(x, \phi_P)$" are co-extensive propositional functions.)

Furthermore a schema analogous to (Sch⁻) could be formulated also in the metalanguage $\mathcal{L}_i$, where predicates regain their rich semantic role:

(Sch⁺) $\ulcorner Ins(a, \phi_P) \urcorner$ is true in $\mathcal{L}_e$ iff $P_\phi a$.

Let us now consider a typical test case for assessing a Realist's proposal about predication: *Bradley's regress about instantiation*. Some philosophers hold that Realism, at least in some versions, falls prey of Bradley's regress about instantiation: indeed, if "*a* is *P*" is true because *a* instantiates *P*-ness, then "*a* instantiates *P*-ness" is true because *a* instantiates the property of instantiating *P*-ness, and so on in an infinite regress. It is remarkable, however, that neither in $\mathcal{L}_i$, nor in $\mathcal{L}_e$, the Bradley's awful regress triggers. To see why, let us prepare the ground by reflecting on the fact that the term "instantiation" is the nominalization of the expression "instantiates", that in turn appears on the scene in the right-hand side of (Sch⁻). But the sole function of (Sch⁻) is to map $\mathcal{L}_i$'s true atomic sentences into the $\mathcal{L}_e$ framework: any pair of substitution instances of (Elementary-P) and (Property-I) is constituted by *different statements*, belonging to *different languages*, but expressing *the same ontological commitment*. (This does not amount to holding that elementary predications and property-instantiation statements are synonymous, or that their difference is "simply a matter of stylistic variation".[20]) And once the trade-off between *connotata* and *nominata* is carried out via (Sch⁻), it is wrong-headed to ask such questions as:

Q1.   Does the predicate "instantiates" pick out something?
Q2.   What binds a property to its bearer(s)?

The short answer to Q1 is: *No*. The expression "*Ins*" that stands in for "instantiates" occurs in a sentence that is a well-formed formula of a language, $\mathcal{L}_e$, where ontological commitments are carried only by singular terms. That predicate is nothing but a technical device required to render in $\mathcal{L}_e$ the ontological commitment of an elementary predication (involving a property-connoting predicate) that, in turn, is a well-formed formula of a language, $\mathcal{L}_i$, in which ontological commitments are harbored by both singular terms and predicates, and can be expressed by both first-order and second-order quantifiers. Therefore, if the predicate "instantiates" occurs in a sentence belonging to a language where ontological commitments are carried only by singular terms, it doesn't pick out anything, and *a fortiori* it doesn't connote something like *instantiation* – whether it is conceived as a *relation*, a *non-relational tie*, or as any other unifying-/binding-entity. The very formulation of Q1 is misleading in as much as it means:

Q1'.   From "$Ins(a, \phi_P)$" can we derive "$\exists X\, X(a, \phi_P)$"?

And, of course, the answer is simply in the negative, since second-order quantification is not allowed in $\mathcal{L}_e$. Moreover, the schema (Sch⁻) is not reiterative, that is, it would be wrong to apply it to the result of any of its previous applications, for it

---

20 The synonymy thesis is held, among others, by Ramsey 1925, Strawson 1974, and Quine 1980. For a criticism of the synonymy thesis see Schnieder 2006.

takes as inputs names of sentences that are true in $\mathcal{L}_i$, and outputs ontologically equivalent $\mathcal{L}_e$-sentences. Since no $\mathcal{L}_e$-sentence is an $\mathcal{L}_i$-sentence, (Sch⁻) cannot bring "*Ins*" nominalization about: hence Bradley's regress is escaped.

Some realists would reply that if this is true, then the (allegedly) compelling question *Q2* remains unanswered, and the kind of Platonism I prefer would as a result be an Ostrich Platonism. But – once again – *Q2* is a loaded question in as much as it comes with the presupposition that we have just questioned and rejected as false: *Q2* presupposes that there is something more to be deduced from a property-instantiation sentence than just the existence of the *correlata* of its singular terms. But for any given property-instantiation sentence, the distinction between its referential and non-referential positions is marked by the distinction between individual constants and the purely extensional predicate constants: thus "*Ins*", being a purely extensional predicate constant, is ontologically uncommitting.

Things are the same in $\mathcal{L}_i$, as it is a language that simply lacks purely extensional predicate constants; therefore $\mathcal{L}_i$ is a language into which "*Ins*" is missing: therein no predicate is purely extensional, and there is no need to introduce an *ad hoc* expression that ties predicatively an *n*-adic property-connoting predicate to *n* individual constants as any property-connoting predicate does this by itself. This is why *Q1* is wrong-headed even when formulated within $\mathcal{L}_i$. The same holds for *Q2*: in asking what is that something that binds a property to its bearer(s), it presupposes that *there is* something that performs this task. But how can we ever infer this? Eventually *Q2* does not make any sense as in $\mathcal{L}_i$ the statement "$\exists X\, X(a, \phi_P)$" cannot be derived from an elementary predication such as "$P_\phi a$".

6 Concluding remarks

Let us now return to the three issues that arise from Armstrong's arguments against Ostrich Nominalism, that is the One over Many question, the question about predicates' ontological seriousness, and the truthmaker question. I will briefly illustrate how the version of Platonism I prefer can solve such issues. We know that the One over Many question asks how two (or more) numerically distinct things, say *a* and *b*, can both be *P* – say white. The Platonist's answer is the same as the one provided by the Ostrich Nominalist: *a* and *b* are both white because *a* is white and *b* is white. (Realists as well as non-Realists often place the One over Many question side by side with the so-called *Many in One question*:

how a white and wise thing, *a*, can be both white and wise?[21] My answer is: Since *a* is white and *a* is wise, *a* is both white and wise.) More generally, according to the version of Platonism I prefer elementary predications need no analysis: *each of them is all right as it is*.[22] The issue raised by Armstrong's harlot argument has to do with the following prescription: *we should take predicates with ontological seriousness* – as Armstrong says. As we have seen so far, in the version of Platonism I prefer predicates are taken with the utmost ontological importance. Finally, concerning the truthmaker question "What is the truthmaker for '*a* is *P*'?", it is important to highlight that it is one thing to look for an *explanation* for the truth of "*a* is *P*"; while it is quite another to look for a *truth-necessitator* for its truth. The version of Platonism I prefer provides a straightforward general explanation for alethic facts: "*a* is *P*" is true because *a* is *P*. But if a truthmaker must also be a truth-necessitator, and if the search for truthmakers leads us to ask such ill-formed questions as *Q1* and *Q2* (see above), then the version of Platonism I prefer is utterly incompatible with the truthmaker-driven approach to predication, in that the latter invites us to search for something that *cannot be found*.

Acknowledgments: I would like to thank Achille C. Varzi for his support and his insightful comments on the topics covered here. I am also very grateful to Francesco Gallina, Anna-Sofia Maurin, Francesco Orilia and William Vallicella for their very constructive comments.

Bibliography

Armstrong, D. M. 1978a. *Universals and Scientific Realism, Vol. 1, Nominalism and Realism*. Cambridge: Cambridge University Press.
Armstrong, D. M. 1978b. *Universals and Scientific Realism, Vol. 2, A Theory of Universals*. Cambridge: Cambridge University Press.
Armstrong, D. M. 1980. "Against 'Ostrich' Nominalism: A Reply to Micheal Devitt". *Pacific Philosophical Quarterly*, 61, 440–449.
Armstrong, D. M. 1989. *Universals: An Opinionated Introduction*. Boulder: Westview Press.
Armstrong, D. M. 2004. *Truth and Truthmakers*. Cambridge: Cambridge University Press.

---

[21] Cf. Campbell 1990, 29, Rodríguez-Pereyra 2000, MacBride 2002, Maurin 2002, 61–4.
[22] Does this mean – once again – that the version of Platonism I prefer is an Ostrich Platonism? In this case I'd say that if to be an ostrich is to believe that the so-called "One over Many problem" is *not* a genuine problem, and that ordinary elementary predications do not stand in need of analysis, then yes: the version of Platonism I prefer can be labeled as "Ostrich Platonism". (Peter van Inwagen has recently accepted such a label for his Platonism as well – cf. van Inwagen 2014, 214–5.)

Bealer, G. 1982. *Quality and Concept*. Oxford: Clarendon Press.
Calemi, F. F. 2012. *Dal nominalismo al platonismo. Il problema degli universali nella filosofia contemporanea*. Milano: Mimesis.
Calemi, F. F. 2014. "The Nominalist's Gambit and The Structure of Predication". *Metaphysica. The International Journal for Ontology and Metaphysics*, 15(2), 313–328.
Campbell, K. 1990. *Abstract Particulars*. Oxford: Blackwell.
Cocchiarella, N. B. 1986. *Logical Investigations of Predication Theory and the Problem of Universals*. Napoli: Bibliopolis.
Künne, W. 2006. "Properties in Abundance". In Strawson, P. F. and Chakrabarti, A. (eds.). *Universals, Concepts and Qualities: New Essays on the Meaning of Predicates*. Aldershot: Ashgate, 249–300.
Loux, M. J. 1978. *Substance and Attribute: A Study in Ontology*. Dordrecht: Reidel.
MacBride, F. 2002. "The Problem of Universals and the Limits of Truth-Making". *Philosophical Papers*, 31(1), 27–37.
Maurin, A.-S. 2002. *If Tropes*. Dordrecht: Kluwer Academic Publishers.
Melia, J. 2005. "Truthmaking without Truthmakers". In Beebee, H. and Dodd, J. (eds.). *Truthmakers: The Contemporary Debate*. New York: Oxford University Press, 67–84.
Mill, J. S. 1843. *System of Logic*. Londra: Routledge.
Oliver, A. 1996. "The Metaphysics of Properties". *Mind*, 105(417), 1–80.
Quine, W. V. O. 1948. "On What There Is". *Review of Metaphysics*, 2(5), 21–38. Reprinted in Quine, W. V. O. 1980. From a Logical Point of View. Cambridge: Harvard University Press, 1–19.
Quine, W. V. O. 1970. *Philosophy of Logic*. Englewood Cliffs: Prentice-Hall.
Ramsey, F. P. 1925. "Universals". *Mind*, 34(136), 401–417.
Rodríguez-Pereyra, G. 2000. "What is the Problem of Universals?". *Mind*, 109(434), 255–273.
Rodríguez-Pereyra, G. 2002. *Resemblance Nominalism: A Solution to the Problem of Universals*. Oxford: Clarendon Press.
Schnieder, B. 2006. "Attributing Properties". *American Philosophical Quarterly*, 43(4), 315–328.
Searle, J. R. 1969. *Speech Acts: An Essay in the Philosophy of Language*. London: Cambridge University Press.
Strawson, P. F. 1974. *Subject and Predicate in Logic and Grammar*. London: Methuen & Co.
Strawson, P. F. 1987. "Concepts and Properties or Predication and Copulation". *The Philosophical Quarterly*, 37(149), 402–406.
Swoyer, C. 1999. "How Ontology Might Be Possible: Explanation and Inference in Metaphysics". *Midwest Studies in Philosophy*, 23(1), 100–131.
Van Inwagen, P. 2014. "Relational vs. constituent ontologies". In van Inwagen, P. 2014. *Existence: Essays in Ontology*. New York: Oxford University Press, 202–220.
Wiggins, D. 1984. "The Sense and Reference of Predicates: A Running Repair to Frege's Doctrine and a Plea for the Copula". *The Philosophical Quarterly*, 34(136), 311–328.
Wolterstorff, N. 1970. *On Universals: An Essay in Ontology*. Chicago: University of Chicago Press.

Peter van Inwagen
In Defense of Transcendent Universals

1 Armstrong's ontological method

David Armstrong and I have very different ideas about method in ontology (and more generally in metaphysics, and, more generally still, in philosophy), and that fact alone is perhaps sufficient to explain why our positions on the questions of the existence and nature of universals are so different – so vastly different.[1]

I will begin by giving a rather abstract description of the method that Armstrong has used in his work on "the problem of universals" – or, as I should prefer to say, the questions 'Are there universals?' and 'Given that there are universals, what is their nature?'. (I put the phrase 'the problem of universals' in scare-quotes and suggest an alternative phrase because, although I believe that there are many philosophical problems about or raised by universals, I do not think that there is any one philosophical problem that deserves to be called the problem of universals.)

Armstrong's approach to these questions is the approach of a theory builder: he approaches them by attempting to construct a theory of universals – an explanatory theory, a theory whose purpose it is to explain certain data. (He will conclude that there are universals because postulating their existence helps to explain these data. He will ascribe a certain nature to them because ascribing that nature to them helps to explain these data.) He begins by collecting and setting out the data the theory is to explain. Here are some examples of facts (I have no objection to describing them as facts) that are among these data:

> The display
>
> | THE | THE |
>
> in one sense contains one word and in another sense contains two words.

> Two tennis balls are before us; they are numerically distinct, and yet they are identical in certain respects: for example, they are identical in color, shape, and size (each is optical yellow; each is a ball; the diameter of each is 6.7 cm).

---

[1] For reasons of both sentiment and stylistic convenience I will speak of David in the present tense.

Peter van Inwagen: University of Notre Dame, email: peter.vaninwagen.1@nd.edu

> The class of dachshunds exhibits greater internal unity than the class of pythons and hummingbirds.

Having collected these (and many other) data, he attempts to construct (or to discover by surveying the relevant philosophical literature) theories that account for them or purport to account for them. He does his best to refine and improve the theories he constructs or discovers, to present each of them in the strongest form possible. He proceeds to compare the members of the inventory of theories so obtained with one another in order to see which of them is the best. He is, however, aware that it is all but inevitable that there will be no straightforward, unassailable answer to the question which of these theories is the best. For one thing, philosophers may disagree about what features a good theory may have or should have. Armstrong himself counts "postulates the existence only of spatio-temporal entities" as a "good making" feature of a theory, but other philosophers will less friendly toward, perhaps even hostile to, theories of universals that postulate only spatio-temporal entities. Or some may see this feature as a great advantage in a theory and others as an advantage but only a very small one. It is, moreover, unlikely that even philosophers who in the abstract accept the same features as advantages and the same features as disadvantages will agree about which of a given class of theories is the best. He has said,

> We have to accept, I think, that straightforward refutation (or proof) of a view in philosophy is rarely possible. What has to be done is to build a case against, or to build a case for, a position. One does this, usually, by examining many different arguments and considerations for and against a position and comparing them with what can be said for or against alternative views. What one should hope to arrive at [...] is something like an intellectual cost-benefit analysis of the views considered.
>
> [...] One important way in which different philosophical and scientific theories about the same topic may be compared is in respect of intellectual economy. In general, the theory that explains the phenomena by means of the least number of entities and principles (in particular, by the least number of *sorts* of entities and principles) is to be preferred.
>
> [...] *Other things being equal*, I shall account the more economical theory the better theory.[2]

My own approach to ontological questions (and, in particular, to the questions of the existence and nature of universals) is entirely different.

As I see matters, the things we say in everyday life and in the sciences, and our everyday and scientific beliefs, have ontological implications, among them implications as regards the existence and nature of universals. It is, I believe, the task of ontology to draw out the ontological implications of our everyday and scientific

---

[2] Armstrong 1989, 19–20. Italics in original. I take 'positions' and 'views' to be – in this context – stylistic variants on 'theories'.

beliefs and assertions. And the way to draw out these implications (I contend) essentially involves the method Quine has recommended. By way of example, consider the following argument.

> Any two mature unmaimed conspecific female spiders have the same anatomical characteristics
>
> Any spider and any insect share certain anatomical characteristics

Therefore

> For any insect and any two mature unmaimed conspecific female spiders, there are anatomical characteristics that belong to that insect and to both spiders.

This argument is valid. If anyone doubts its validity, those doubts can be removed by the simple expedient of pointing out that its "obvious" translation into the quantifier-variable idiom is *formally* valid. And that obvious translation is this or something very much like it:

> $\forall x \forall y$ (x is a mature unmaimed female spider & y is a mature unmaimed female spider & x and y are conspecific $\rightarrow$ $\forall z$ (z is an anatomical characteristic $\rightarrow$. x has z $\leftrightarrow$ y has z)).
>
> $\forall x \forall y$ (x is spider & y is an insect. $\rightarrow$ $\exists z$ (z is an anatomical characteristic & x has z & w has z))

Therefore

> $\forall x \forall y \forall z$ (x is an insect & y is a mature unmaimed female spider & z is a mature unmaimed female spider & y and z are conspecific . $\rightarrow$ $\exists w$ (w is an anatomical characteristic & x has w & y has w & z has w)).[3]

Well and good. But... consider the second premise of this argument; and consider the obvious truth '$\exists xx$ is a spider . & $\exists xx$ is an insect'. From these two sentences one may formally deduce

> $\exists x$ is an anatomical characteristic.

It seems, therefore, that careful formal analysis of some very simple beliefs we all have about spiders and insects shows that these beliefs entail the existence of

---

[3] One might want to insert '& $y \neq z$' at the obvious place. The English is ambiguous. This ambiguity is not relevant to my argument.

anatomical characteristics. Any anatomical characteristic is of course a characteristic. And 'characteristic', 'feature', 'property', 'attribute' and 'quality' are all synonyms or as near to being synonyms as makes no matter. Some of our most ordinary beliefs, therefore, in Quine's famous phrase, carry ontological commitment to properties or attributes. Or, so, at any rate, I have often contended. And properties or attributes are certainly "universals."[4] And, if one is convinced that arguments of this sort prove that we are "committed" by various things that we say and believe (and many of which most of us would regard as obviously true) to such things as "characteristics," one may go on to consider philosophical arguments for and against such theses as 'There are characteristics that nothing has' and 'Some characteristics are impossible' and 'Characteristics are ontological constituents of particulars'. That is, one may go on to investigate the nature of "characteristics."

It is not my intention to discuss arguments of these kinds in any greater detail than this. Those who want detail can find it in my "A Theory of Properties."[5] But if I say that the method I have employed in my attempts to reach conclusions concerning the existence and nature of universals has been, first, to attempt to show that various of our everyday and scientific beliefs and assertions imply the existence of characteristics (or properties or features or qualities or attributes – call them what you will) and, secondly, to investigate the nature of these items, I have not said enough. I must also answer the question whether my approach to the questions 'Do universals exist?' and 'Given that universals exist, what is their nature?' is not as much "the approach of a theory builder" as Armstrong's. After all, is the title of my principal work on the topic not "A Theory of Properties"?

And here is my answer. There is a significant difference between Armstrong's "theory" of universals and my "theory" of properties. Armstrong's theory is in part an explanatory theory, a theory that is intended to explain certain data by postulating universals (and ascribing a certain nature to these postulated entities). My theory of properties is a purely descriptive theory: a theory that consists

---

[4] A "universal" is a thing that has, or can have, instances. They are the entities that are "universal to" their instances. A sentence-type (if sentence-types in fact exist; I omit the corresponding qualification in the examples that follow) is a universal whose instances are its tokens. A novel is a universal whose instances are the tangible copies of that novel. A property is a universal whose instances are the things that have that property. (A terminological warning: in the usage of some philosophers, "property-instances" are *not* the things that I am describing as instances of that property.) A relation is a universal whose instances are the sequences of things that stand in that relation. (Philosophers who affirm the existence of impossible properties and relations will obviously need to qualify the statement 'A universal is a thing that has, or can have, instances' in some way.)

[5] Van Inwagen 2004.

entirely of statements about what sorts of things properties (characteristics, etc.) are and how they are related to one another and to their instances. (Of course any "Armstrong style" theory of universals will have a descriptive component. It will incorporate descriptive statements like 'A universal is wholly located at any place at which one of its instances is located'. But the "descriptive component" of a "van Inwagen style" theory of universals is the whole of the theory.) My method for approaching the question of the *existence* of properties is to attempt to show that our language and our thought (the affirmations and beliefs we bring to philosophy) define a role that can plausibly be described as the "property role," and to defend the thesis that there are objects that play that role. My *theory* of properties is an attempt to describe the nature the objects must have if they are to play that role.

My theory of the nature of the things that play the property role – for example, the things, other than spiders and insects, that we "quantify over" when we assert that spiders and insects share certain anatomical characteristics – has been aptly described by Kenny Boyce as "lightweight platonism."[6] 'Platonism' because the theory contends that universals are invisible, intangible, achronic and non-spatial things that exist at all times and in all places[7] and in all possible worlds, and whose existence is serenely indifferent to whether they have instances: if a universal has instances/has no instances at one time and has no instances/has instances at a later time, that is a "mere Cambridge" change in that universal; there is no sense whatever in which a universal is a "component" or "constituent" of its instances. 'Lightweight' because, as I view matters, universals are "anetiological"; they have no causal powers and they are not such as to respond to the causal powers[8] of those things that do have causal powers. (Plato's Forms or Ideas certainly had causal powers.)

---

[6] Kenneth A. Boyce, *Towards a Fictionalist Nominalism* (A dissertation submitted to the Graduate School of the University of Notre Dame in 2013 in partial fulfillment of the requirements for the degree of Doctor of Philosophy.) See pp. 31*ff*.

[7] If they are "achronic" and "non-spatial," what can be meant by saying that they exist at all times and in all places? Well, let us say that if God were to provide us at some time and place with a list of all universals, he would have provided the same list at any other time or place. Perhaps, rather than saying that universals exist at all times and in all places, it would be better to say that statements about the existence of universals – unlike statements about their having or not having instances – cannot be temporally or spatially qualified. To ask when or where a universal exists is to ask a question that betrays a misunderstanding of the nature of universals.

[8] As I see causal powers, they are themselves properties; that is, some properties are causal powers. Suppose that a poker that has been removed from the fire is now hot enough to scorch paper. The property "being hot enough to scorch paper" is a causal power but *has* no causal powers; it is concrete particulars (or substances) like the poker that *have* causal powers.

I will not attempt to explain why I maintain that properties (the only universals I have discussed in any detail) have the features listed in the preceding paragraph.[9] My purpose in this paper is to attempt to respond to Armstrong's criticism of a class of theories of universals of which mine is certainly a member. He calls the doctrine that is common to all such theories transcendent realism. (Armstrong supposes that transcendent realism is an explanatory theory – that the platonic or transcendent universals whose existence is affirmed by transcendent realism are entities whose existence transcendent realists postulate to explain the same data as those he is attempting to explain.[10] As we have seen, I do not "postulate" the existence of transcendent universals – I have no particular objection to calling the "properties" of which my theory treats 'transcendent universals' –, and my theory is in no sense an explanatory theory. I am going to treat Armstrong's supposition as a false statement about the reasons a transcendent realist must have for being a transcendent realist, and not as a part of the content of transcendent realism. And, indeed, the explicit definition of 'transcendent realism' that Armstrong provides says nothing about the reasons a transcendent realist might have for embracing transcendent realism.)

2 Armstrong's primary critique of transcendent realism

I will now set out Armstrong's account of transcendent realism and what I believe to be his primary critique of this position.[11] In the course of doing this, I will show that his critique does not apply to my own version of transcendent realism.

We have, first, from the "Glossary of terms used and principles formulated" at the end of the first volume of *Universals and Scientific Realism*:

> *Immanent realism*: the doctrine that admits universals but denies that they are transcendent. (p. 137)

---

9 The interested reader can consult van Inwagen 2004, particularly pp. 125–138.
10 "Transcendent [Universals] . . . are theoretical entities, standing apart from the ordinary world, postulated in the same general sort of way that atoms or genes were postulated, to explain certain phenomena. In a matter of fundamental ontology, this is a *prima facie*, though certainly no more than a prima facie, disadvantage of the theory of [Transcendent Universals]. . . . Not only are [Transcendent Universals] postulated entities, but so is the relation between particular and [Transcendent Universal] which bestows a nature upon the particular." (Armstrong 1978, 66).
11 The quotations that follow are from *Universals and Scientific Realism: Volume I*.

> *Transcendent realism*: The doctrine that universals exist separated from particulars. (p. 140)

Chapter 7 of that volume, "Transcendent Universals," opens with these words:

> We turn now to a version of Realism[12] [It is] a *Relational* theory. According to this view, *a* has the property, F, if and only if *a* has a suitable relation to the transcendent universal or Form of F... It will be convenient to speak of transcendent universals as Forms. (p. 64)

These statements do not apply to my own version of transcendent realism. As I see matters, "the property F" *is* a transcendent universal. All properties – I contend – are transcendent universals, and some at least of transcendent universals are properties. I am, of course, willing to say that *a* has the property F if and only if *a* has a suitable relation to the property F,[13] but I take that "suitable relation" to be the one expressed by the verb 'has' in the previous clause. That is, I am willing to say, "*a* has the property F if and only if *a* has the property F" (as who wouldn't be?).

And what is this relation I call (and all other speakers of English call) 'having'?[14] 'Has' is not a technical term, but a word that belongs to our ordinary speech. Of course, like most everyday words, it has many senses, and these must be distinguished in a philosophical discussion of the nature of universals (as the 'true' of 'true friend' and the 'true' of 'true statement' must be distinguished in a philosophical discussion of the nature of truth). Obviously, when we say that a thing has a certain property, we are not using 'has' in any of the senses of the word illustrated by the following examples: 'Alice has a cold/has a grievance /has a Lexus/has a husband'. 'Has', like 'can' and 'is' is a very versatile word. But the sense of 'had' in the sentence 'Solomon had the property wisdom' (and in any sentence with the same semantical structure) is uniquely determined by, and is evident from an examination of, the sentence as a whole. The proposition expressed by this sentence, moreover, can be expressed by other sentences, sentences that make use of other everyday idioms. The following sentences of ordinary English all express the same proposition, or at least come as close to

---

12 The "version of Realism" being transcendent realism.
13 Although I should prefer to say 'if and only if *a* stands in a suitable relation to the property F' or 'if and only if *a* and the property F are suitably related'.
14 *I* call it a relation. Others prefer other terms – 'tie' being one popular alternative. But call it what you will. Note that "having," if it is a relation, would seem to be an *internal* relation (like "is the same color as"), at least when it is asserted to hold between an object and an *intrinsic* property. Whether "having" holds between Solomon and wisdom – assuming that wisdom is an intrinsic property – is "settled" by the intrinsic properties of Solomon and wisdom.

expressing the same proposition as do 'The capital city of France is Paris' and 'Paris is the French capital':

> Solomon had the property wisdom.[15]
>
> Wisdom was one of Solomon's properties.
>
> The property wisdom belonged to Solomon.
>
> Wisdom can be truly predicated of (or 'ascribed to') Solomon.

This relation, I contend, the relation these sentences assert to hold between Solomon and wisdom, is as familiar and well-understood as it is hard to explain or give an account of. I would say that it was hard to explain because it was hard – impossible, in fact – to find simpler or better-understood ideas in terms of which to explain it.

Now in what sense is transcendent realism a "relational" theory? Not, surely, because it implies that Solomon had the property wisdom if and only if Solomon had the property wisdom? Perhaps what Armstrong was trying to say in terms of quantification over properties and "transcendent universals" could be put more clearly in terms of quantification over terms (names and definite descriptions) and predicates:

> Transcendent realism is a Relational theory. According to this view, for any term a and any predicate **F**, the sentence ⌜a is **F**⌝ if and only if a has the property of being **F**⌝ expresses a true proposition (and so for 'was' and 'had', *mutatis mutandis*) – and, indeed, a *necessarily* true proposition.[16]

For example, on this account of transcend realism, the sentence 'Solomon was wise if and only if Solomon had the property of being wise' expresses a (necessarily) true proposition. And this (so I interpret Armstrong's reasons for calling tran-

---

[15] I concede that it is unlikely that anyone but a philosopher would ever utter or inscribe the sentence 'Solomon had the property wisdom'. (Presumably for pragmatic reasons: because in any ordinary context, anyone who said, "Solomon had the property wisdom," would set his or her audience to wondering what the point of saying that rather than "Solomon was wise" might have been.) But an utterance of the following sentence would be unsurprising: 'Wisdom is not a usual property of hereditary monarchs; if Solomon indeed had this property, as pious tradition holds, he was one of very few kings who did'.

[16] Why do I formulate this thesis by employing quantifiers that bind nominal variables ranging over linguistic items (terms and predicates)? Why do I not use a quantifier that binds variables in predicative positions; why do I not use an expression along the lines of '$\forall x \forall y (Fx \leftrightarrow x$ has the property of being **F**)'? The short answer is that I believe that the idea of variables that occupy non-nominal positions (e.g., predicative positions and sentential positions) makes no sense.

scendent realism a relational theory) implies that the proposition that Solomon was wise is a *relational* proposition owing to the fact that, according to transcendent realism, it is true if and only if Solomon bears a certain relation – "having" – to the property of being wise (or more idiomatically, to the property wisdom). And this is not simply a contingent truth if transcendent realism is true: transcendent realism implies that Solomon could not *possibly* have been wise without bearing this relation to the transcendent universal wisdom, an object that exists independently of Solomon and whose being in no way overlaps his being (an object that is, as one says, an inhabitant of the Platonic heaven).[17]

If that is indeed a fair reconstruction of Armstrong's reason for saying that transcendent realism is a relational theory,[18] it is not a good reason. There are at least two points to be made against it. I will conclude this section with a discussion of the first of these points, a very minor one. The second and much more important one will be addressed in section 3. The first point, the minor point, , is that the universal quantification

> For any term *a* and any predicate **F**, the sentence ⌜*a* is **F** if and only if *a* has the property of being **F**⌝ expresses a true proposition

is not true – or at least it is an eminently defensible position that it is not true. For consider wisdom. It is evident that this property is not itself wise; we may therefore say that wisdom is *non-self-applicable* (unlike, say, the property "being an abstract object" – which *is* an abstract object). And a good case can be made for saying that the sentence

> Wisdom is non-self-applicable if and only if wisdom has the property of being non-self-applicable

expresses a false proposition – or, at any rate, does not express a true proposition. The left-hand constituent of the biconditional seems obviously true, and it is hard to see how its right-hand constituent could be true, owing to the fact that (as Russell has shown) it is hard to see how there could be such a property as the property of being non-self-applicable. (Let 'N' abbreviate 'the property of being non-self-applicable'. If N exists, N is non-self-applicable if it is self-applicable

---

[17] This last qualification is important. Without it, Armstrong's own theory of universals could with equal justification be described as a relational theory. For anyone who accepts Armstrong's theory will accept the following thesis: For any term *a* and any predicate **F**, ⌜A thick particular *a* is **F** if and only if *a* has the property of being **F** as a constituent⌝ expresses a true proposition.

[18] Paul Audi clearly and explicitly gives this as a reason for classifying my own theory of properties as a relational theory. See Audi 2013, 755–756.

and self-applicable if it is non-self applicable. N is therefore – if N exists – both self-applicable and non-self-applicable. Hence, N does *not* exist.) And if 'the property of being non-self-applicable' has no referent, then the right-hand constituent of the biconditional is false and the biconditional is therefore false. (At any rate, the biconditional is certainly not *true*.)

Armstrong's argument for the conclusion that transcendent realism is a relational theory does not really require that transcendent realists accept this general principle, however. The argument requires only a restricted version of the general principle:

> For any term *a* and any predicate **F**, if *a* denotes a concrete particular, then the sentence ⌐*a* is **F** if and only if *a* has the property of being **F**⌐ expresses a true proposition

And it seems evident that any transcendent realist must accept this restricted principle.

If transcendent realism indeed implies that the proposition that Solomon was wise is a relational proposition, then it is easy to see why one should not be a transcendent realist: that proposition is obviously not a relational proposition; for no object *x* and no relation R is that proposition the proposition that R holds between Solomon and *x*. The only object that in any way "figures in" the proposition that Solomon was wise is Solomon. That proposition says of that object, Solomon, that it, or he, is wise, but in saying that of him, it does not affirm that there is some other object, some object that exists independently of him (even if that object be one that is in some way logically or semantically or ontologically intimately connected with the predicate 'wise'), to which he bears some relation – "having" or any other relation.

3 A reply to the primary critique

I will show that transcendent realism does *not* imply that propositions expressed by sentences of the form 'S is **F**' (where 'S' represents a term denoting a concrete particular and '**F**' represents a predicate expressing an intrinsic property) are relational propositions. (Let us call such propositions SIP propositions—for 'subject-intrinsic predicate'.)

Let us consider the argument I have ascribed to Armstrong for the thesis that transcendent realism implies that all SIP propositions are relational propositions. Like Gaunilon *vis-à-vis* St Anselm's argument, I will attempt to convince my readers that there has to be something wrong with the argument by presenting a parody. (In fact, I will generously present two parodies.)

The first parody:

Propositionalism (the thesis that the bearers of truth-value are language-independent abstract objects) implies that all SIP propositions are relational propositions. Consider, for example, the proposition that Solomon was wise, a representative SIP proposition. Propositionalism implies that the proposition that Solomon was wise is a relational proposition owing to the fact that, according to propositionalism, this proposition is true if and only if Solomon bore a certain relation to the proposition that Solomon was wise (the relation "being such that it is true," the relation expressed by the open sentence 'x is such that y is true'). And this is not simply a contingent truth if propositionalism is true: propositionalism implies that Solomon could not possibly have been wise without having borne this relation to the proposition that Solomon was wise, an abstract object that exists independently of Solomon and whose being in no way overlaps his being (an object that is, as one says, an inhabitant of the Platonic heaven).[19]

The second parody:

Arithmetical realism (the thesis that numbers are real objects) implies that some SIP propositions are relational propositions. Consider, for example, the proposition that Solomon was bipedal, which is certainly an SIP proposition. Arithmetical realism implies that the proposition that Solomon was bipedal is a relational proposition owing to the fact that, according to arithmetical realism, this proposition is true if and only if Solomon bore a certain relation to the number 2 (the relation "having had legs that were numbered by," the relation expressed by the open sentence 'x's legs were numbered by y'). And this is not simply a contingent truth if arithmetical realism is true: arithmetical realism implies that Solomon could not *possibly* have been bipedal without having borne this relation to the number 2, an abstract object that exists independently of Solomon and whose being in no way overlaps his being (an object that is, as one says, an inhabitant of the Platonic heaven).

---

19 Will someone deny that the being of the proposition that Solomon was wise "in no way overlaps" Solomon's being owing to the fact that Solomon is a constituent of that proposition? Any who takes that position may substitute some "purely qualitative" definite description that denotes Solomon for 'Solomon' in the first parody. 'The ancient king famed for his wisdom who constructed a great temple', for example.

I hope that the lesson of the two parodies is evident. Evident or not, it is this:

> If p and q are metaphysically equivalent propositions (if p and q are true in the same possible worlds) and if p is a relational proposition, it does not follow that q is a relational proposition.

That transcendent realism implies the necessary truth of biconditional propositions like

> Solomon was wise if and only if Solomon had the property of being wise

is therefore not sufficient for classifying transcendent realism as a "relational" theory. That is to say, from the (true) premise that transcendent realism implies that such biconditionals are necessary truths, one may not validly deduce the conclusion that transcendent realism also implies that SIP propositions are relational propositions. To reach the conclusion that transcendent realism is a relational theory, one would need a stronger premise. One would need a premise that implies that a relation more intimate than mere necessary coextensiveness holds between pairs of predicates like 'is wise' and 'has the property of being wise'. One would, in fact, need the following premise or something very much to the same purpose:

> For any term a and any predicate **F**, if a denotes a concrete particular that belongs to the extension of **F**, then the sentence ⌜What it is for a to be **F** is for it to have the property of being **F**⌝[20] expresses a true proposition

One would need a premise that has consequences like this one (given that Solomon belonged to the extension of 'wise'):

> What it was for Solomon to be wise was for Solomon to have the property of being wise.

---

20 Some alternatives: ⌜a's being **F** consists in a's having the property of being **F**⌝; ⌜a's being **F** is grounded in a's having the property of being **F**⌝; ⌜a is **F** in virtue of a's having the property of being **F**⌝; ⌜a is **F** because a has the property of being **F**⌝. Consider the last of these alternatives. Although 'because' is a tricky word, I think that it is closer to the truth to say that the Taj Mahal has the property of being white because it is white than it is to say that the Taj Mahal is white because it has the property of being white. Aristotle has famously said, "It is not because we think truly that you are white [= pale], that you are white, but because you are white we who say this have the truth." (*Metaph.* IX, 1051b). Similarly, one might say, "It is not because whiteness is truly predicable of you that you are white, but because you are white whiteness is truly predicable of you."

Now no doubt there are and have been many transcendent realists who would accept the thesis that what it was for Solomon to be wise was for him to have had the property of being wise. But this thesis is not essential to transcendent realism. I myself reject it. In my view, to say that Solomon's having the property wisdom is what it was for Solomon to be wise is as absurd as saying that what it was for Solomon to be wise is for the proposition that Solomon was wise to be true – or as saying that what it was for Solomon to be bipedal is for the number 2 to have been the number of his legs. In fact, in my view, the only sense the question 'What is it for someone to be wise?' can have is the sense that this question had when it was discussed by Socrates and his companions in dialectic. Since I do not know the answer to that ancient and profound question, I will not use 'wise' and wisdom in the examples in the paragraphs that follow. The examples in the sequel will pertain to the predicate 'white' and whiteness or the property of being white –, a property of physical objects, the common property of a sheet of white paper, a whitewashed fence, and the Taj Mahal.

What is it for an object to be white? I can think of two ways to answer this question. (Both the answers I shall provide are answers I have produced without much reflection. No doubt they could be improved. The second could certainly be expanded.) There is, first, a philosophical answer (but the question to which it is an answer belongs to the philosophy of perception, not to ontology):

> For an object to be white is for it to have a disposition to cause visual sensations of whiteness in normal human observers who view it in ideal circumstances.

And there is, secondly, a physical answer:

> For an object to be white is for it to have a surface that reflects all the component frequencies of white light (light composed of an equal mixture of all the visible frequencies of electromagnetic radiation; the visible component frequencies of sunlight are a good approximation) more or less equally.

Now while Armstrong would not deny that these two "answers" are correct answers to the question 'What is it for an object to be white?' on two perfectly reasonable interpretations of that question, he would maintain that they are not answers to the question *metaphysicians* are asking when they ask, "What is it for an object to be white?" He would perhaps tell us that an object's having a surface that reflects all the components of white light more or less equally poses exactly the same metaphysical problem as the problem posed by an object's being white. The questions 'What is it for an object to be white?' and 'What is it for an object to have a surface that reflects all the component frequencies of white light more or less equally?' (not to mention the question, 'What is it for an object to be disposed to cause visual sensations of whiteness in normal human observers who view it in ideal circumstances?') are equally good and equally difficult metaphysical questions. The answer to the latter question according to transcendent realism (as Armstrong conceives transcendent realism) is

> For an object to have a surface that reflects all the components of white light more or less equally is for it to have a suitable relation to the transcendent universal or Form "having a surface that reflects all the components of white light more or less equally" (or, as I should prefer to say, '… for it to have the property of having a surface that reflects all the components of white light more or less equally').

As I have conceded, some transcendent realists may endorse the statement 'For an object to be white is for it to have the property whiteness'. No doubt these realists would also accept this second statement. I reject both statements. In my view, the question 'What is it for an object to be white?' is a question to be answered either by physics or by a philosophical analysis of color-predicates, depending on the kind of answer one is looking for. And the question, 'What is it for an object to have a surface that reflects all the components of white light more or less equally?' is a question to be answered by physics alone. There simply are no such questions as the supposed metaphysical questions 'What is it for an object to be white?' and 'What is it for an object to have a surface (etc.)?' And, in my view, the declarative sentences that are supposed to be their answers (sentences like 'For an object to be white is for it to bear the relation "having" to a certain object in Plato's heaven, to wit, property of being white' and 'For an object to be white is for it to have the universal *whiteness* as an ontological constituent') are meaningless. (Or perhaps false *a priori*. Compare that postwar-Oxford chestnut 'Quadruplicity drinks procrastination'. It that sentence meaningless or false *a priori*? A case could be made for either answer.)

4 "How can distinct particulars have the same properties?"

I pass now to a second but related characterization of transcendent realism (at least it is a characterization of transcendent realism if the phrase 'the Platonic theory of Forms' in the quotation that follows is intended to refer to transcendent realism). Armstrong has said,

> The problem of universals is the problem how different particulars can nevertheless have the very same properties and relations. It is the problem of generic identity. The Platonic theory of Forms is intended to solve this problem [...] (p. 64)

(He will go on to contend that the "Platonic theory of Forms" does not provide a satisfactory solution to this problem.) But is there any such problem? I will try to explain why – whatever other transcendent realists may think – I think that there is not.

Let A and B be two of the component planks of the fence Tom Sawyer tricked Ben Rogers into whitewashing. Miriam the metaphysician asks, "How can A and B, which are numerically diverse, both be white?" If Miriam addressed that question to me, I could think of only two ways to answer it.

There is, first, the efficient-causal answer:

> Ben Rogers whitewashed A (thus causing it to be white) and Ben Rogers whitewashed B (thus causing it to be white). And, obviously, an agent can whitewash more than one plank.

(Compare: 'How can Alexandria Ariana and Alexandria Asiana, which were numerically diverse cities, both have had the same founder?' 'They were both founded by Alexander the Great; obviously a person can found more than one city'.)

There is, secondly, the formal-causal answer:

> A and B are both such that, if white light falls upon them, they reflect all the component frequencies of that light more or less equally. Obviously, two planks that are physically very similar will reflect light in similar ways.[21]

These two answers can be generalized:

> Two physical objects can both be white because the agencies that can cause a physical object to become white can operate on more than one such object.

> A physical object is white if, when white light falls upon it, it reflects all the component frequencies of that light more or less equally; it is therefore

---

[21] Or perhaps I should say that this is one of two formal-causal answers, the other being, 'A and B both have dispositions to cause visual sensations of whiteness in normal human observers who view them in ideal circumstances. Obviously two different objects that are physically very similar will have the same causal dispositions'. I will not discuss this answer in the text. What I would say if I did discuss it can easily be inferred from what I say about the "physical" formal-causal answer.

possible for two objects to be white because it is possible for more than one object to be such that when white light falls upon it, it reflects all the component frequencies of that light more or less equally.

But neither of these general statements, true though they be, is an answer to the *metaphysical* question our metaphysician Miriam would be asking if she uttered the sentence 'How can two physical objects, which are numerically diverse, both be white? For suppose that one is, like Miriam, convinced that this sentence can be used to ask a question whose answer is neither a causal explanation of the whiteness of both objects nor an account of the physical features common to the two objects that "underlie" whiteness; suppose one believes that this sentence can be used to ask a question whose answer would be an *ontological analysis* of a thing's being white. (A proponent of Platonic "transcendent universals" would give one sort of ontological analysis of a thing's being white; a proponent of Aristotelian "immanent universals" another sort of ontological analysis; a nominalist a third sort of ontological analysis.) Anyone who is convinced that the sentence 'How can two physical objects, which are numerically diverse, both be white?' has these features will certainly have the corresponding beliefs about the interrogative sentences

> How can two physical objects, which are numerically diverse, be such that the agencies that can cause a physical object to become white can operate on both of them?
>
> How can two physical objects, which are numerically diverse, both be such that when white light falls upon them, each reflects all the component frequencies of that light more or less equally?

As Miriam sees matters, the sentence 'How can two physical objects, which are numerically diverse, both be white?' is no more than an example of a question whose answer would be a certain sort of ontological analysis. The fact that the predicate that figured in the question happened to be a color-predicate is irrelevant to the point of the example. The two offset sentences would both be equally good examples of questions of that sort (although the complexity of the predicates that figure in the two sentences perhaps makes them less than ideal as illustrative examples). Thus, neither of the two answers to 'How can two physical objects, which are numerically diverse, both be white?' that I have proposed is relevant to the question Miriam means to be asking.

It will perhaps be evident from what I said in the preceding section that I do not believe that there is any such question as the question Miriam means to be asking. "The riddle does not exist. If a question can be framed at all, then it is also possible to answer it." Miriam's question is non-existent because it cannot be answered; it cannot be answered (even by God) because no set of statements among all possible sets of statements is an answer to it.[22] Certainly, this statement is not an answer to it.

> Two numerically diverse objects can both be white because they can both have the property whiteness (an attribute or quality, an abstract object, a Platonic universal, a transcendent universal, a universal *ante res*).

I do not deny that two objects can both have the property whiteness, and I do not deny that whiteness is all the things listed in the parenthesis. (After all, I am a transcendent realist, and these things are exactly the things we transcendent realists believe.) What I deny is the "because." As an *explanation* the offset sentence is simply absurd – as absurd as

> Two numerically diverse objects x and y can both be white because the proposition that x is white and the proposition that y is white can both be true

and

> Two numerically diverse animals x and y can both be bipeds because it is possible for the number 2 to be the number of x's legs and also to be the number of y's legs.

The general point I want to make is this. A transcendent realist is a metaphysician who believes that universals exist, that these universals exist *ante res* (independently of their instances), that universals are necessarily existent things, that universals are in no sense constituents of particulars, that the only constituents of particulars are their *proper parts* (other "smaller" particulars; for example, electrons, submicroscopic particulars, are "constituents" of macroscopic particulars like cats and canaries),[23] that universals are abstract objects and thus do not and cannot enter into causal relations "from either end": they can be neither agents nor patients.

---

[22] Cf. my discussion of Miriam's colleague Alice in van Inwagen 2011, 389–405.
[23] Thus, transcendent realists reject Armstrong's "layer-cake" model of particulars; transcendent realists agree with nominalists about the nature of particulars: to use Armstrong's dyslogistic term, they are "blobs." (A singularly inappropriate term. To paraphrase a point George Bealer once made in correspondence: if one holds that physically complex particulars like bicycles do not have immanent universals as constituents, it is at best very misleading to say that one therefore regards bicycles as blobs.)

This is the position of the transcendent realists. And why do transcendent realists hold this position? I can think of two reasons. One is the reason that Armstrong supposes must be the reason of all transcendent realists, a reason that parallels his reason for being an immanent realist: there are certain data, certain phenomena, to be explained, and the transcendent realist postulates the existence of transcendent universals to explain these phenomena. The other is my reason: there are good philosophical arguments (arguments that are most emphatically not "inferences to the best explanation") for the existence of immanent universals. (See, for example, the "spiders and insects" argument in section 1.)

5 Arguments, not explanations

If, therefore, one argues that immanent realism is to be preferred to transcendent realism because immanent realism better explains certain data (e.g., that objects that are numerically diverse can nevertheless be identical in a certain respect), that is in my view no argument at all – for there is and could be no such thing as an explanation of those data. Or, more precisely, there is no such thing as an explanation of those data that is of the sort the immanent realist claims to provide or of the sort that transcendent realists who have rival explanations of them claim to provide: explanations at whose core lies an ontological analysis of being wise, of being white, of being bipedal, of being... (and so on). And this is because there is and could be no such thing as an ontological analysis of being wise or being white or being bipedal.[24] There are indeed various necessarily true and non-trivial statements (some of them are instances of general theses of considerable metaphysical significance) of the forms

$\forall x(x$ is wise $\leftrightarrow ...x...)$

$\forall x(x$ is white $\leftrightarrow ...x...)$

$\forall x(x$ is bipedal $\leftrightarrow ...x...)$

but none of these statements is an ontological analysis of being wise (white, bipedal) and none of them be used to provide an explanation of what it is for a thing to be wise (white, bipedal).

---

[24] There are in my view, excellent arguments for the falsity of nominalism (the thesis that there are no universals of any kind). But this is not one of them (despite its being a true statement): Nominalism can provide no explanation of the fact that objects that are numerically diverse can nevertheless be identical in a certain respect.

Similarly, there are various statements of the form

$\exists x \exists y(x \neq y$ & x is white & y is white$)$ & $\forall x \forall y(x$ is white & y is white $. \leftrightarrow \ldots x \ldots y \ldots)$

whose second conjunct is necessarily true and non-trivial (and so for any other predicate). But none of these statements explains how it is possible for objects that are numerically diverse to be (nevertheless) identical in a certain respect.

In sum, arguments whose conclusions pertain to the existence and nature of universals can never be inferences to the best explanation. But there are plenty of arguments for theses that pertain to universals that are not inferences to the best explanation. I will close with an example of such an argument. (I do not present this argument to establish its conclusion, but only as an example of an argument for a significant thesis about universals that is not an inference to the best explanation, is not an argument of the general form 'We should accept the thesis that universals have the property **F** because ascribing **F** to them enables us better to explain the following data...'. Because I offer this argument that follows only as an example of an argument of a certain kind, I will not consider the objections that might be raised against it – some of which are formidable.)

One extremely important difference between immanent realism and transcendent realism is that immanent realists deny that there can be uninstantiated properties or attributes and transcendent realists affirm the possibility of uninstantiated properties. The following argument for the possibility of whiteness existing uninstantiated is therefore an argument for the falsity of immanent realism.)

> Consider whiteness or the property of being white. (The immanent realist of course believes that there is such a property.) Suppose for *reductio* that it is impossible for this property, whiteness, to exist uninstantiated – that is, to exist if nothing has it. There are obviously possible worlds in which nothing has the property whiteness. It follows that there are possible worlds in which whiteness does not exist. And the proposition that something is white exists only if the proposition that something has the property whiteness exists. There are, therefore, worlds in which the proposition that something is white does not exist. Let *w* be any such world. An object (of any sort, in any logical or metaphysical category) can have a given property only if that object exists. It is therefore not the case that, in *w*, the proposition that something is white has the property *possible truth*. And, therefore, it is also not the case that, in *w*, the proposition that something is white is possibly true. The world *w* is accessible from the actual world (we have, after all, described it in terms of what *is* possible). But the actual world is not accessible from *w*, since in the actual world the proposition that something is

white is true and it is not the case that in *w* the proposition that something is white is possibly true: there is a proposition that is not possibly true in *w* and is true in the actual world. Hence, the accessibility relation is not symmetrical. But it is absurd to deny that the accessibility relation is symmetrical, and the *reductio* of the thesis that it is impossible for whiteness to exist uninstantiated is complete. *Hence*: It is possible for whiteness to exist uninstantiated—and immanent realism is therefore false.

Bibliography

Armstrong, D. M. 1978a. *Universals and Scientific Realism, Vol. 1, Nominalism and Realism*. Cambridge: Cambridge University Press.

Armstrong, D. M. 1989. *Universals: An Opinionated Introduction*. Boulder: Westview Press.

Audi, P. 2013. "How to Rule Out Disjunctive Properties". *Noûs*, 47(4), 748–766.

Boyce, K. A. 2013. *Towards a Fictionalist Nominalism* (A dissertation submitted to the Graduate School of the University of Notre Dame in 2013 in partial fulfillment of the requirements for the degree of Doctor of Philosophy.)

Van Inwagen, P. 2004. "A Theory of Properties". In Zimmerman, D. (ed.), *Oxford Studies in Metaphysics, Vol. 1*. Oxford: Oxford University Press, 107–138. Reprinted in van Inwagen, P. 2014. *Existence: Essays in Ontology*. Cambridge: Cambridge University Press, 153–182.

Van Inwagen 2011. "Relational vs. Constituent Ontologies". *Philosophical Perspectives*, 25, 389–405. Reprinted in Reprinted in van Inwagen, P. 2014. *Existence: Essays in Ontology*. Cambridge: Cambridge University Press, 202–220.

Peter Simons
Armstrong and Tropes

> So the philosophy of tropes is riding high.
> David Armstrong, *U* 125.[1]

Appreciation

In 1974, while a graduate student at the University of Manchester, I first heard David Armstrong give a talk. It was on various regress arguments against nominalism, later published in *NR*. At that time I was not a nominalist and the arguments seemed sound. What impressed me much more forcibly however was Armstrong's refreshing directness in addressing metaphysical issues. At that time, metaphysics was largely still under the domination of the philosophy of language, and arguments about the nature of universals tended to go via consideration of predicates, semantics and so on. Armstrong's rejection of bad old arguments from meaning cut through that tangle like Alexander's sword through the Gordian knot, and we were left face to face with the metaphysical question itself: are there universals, or are there not? Just as in the phenomenological tradition Roman Ingarden had broken away from Husserl's transcendental anxietizing, so in the analytical tradition Armstrong broke away from Strawson's Aristotelian-Kantian linguistic metaphysics-lite and Quine's Carnapian insouciance as to ultimates. The idea that metaphysics might regain her status as the Queen of Philosophy was implanted, and has stayed with me ever since. I owe David a great intellectual debt for making this clear by living example.

The next time I saw David was in 1990 at Zinal, by which time we and things had both moved on, and this essay is about some of those things. In the meantime I had become a fairly convinced nominalist, but that disagreement aside, we found we had much in common, not least resembling passions for history: political, military and especially of course naval. Knowing his conservative monarchism did not chime with my liberal republicanism, I steered clear of contemporary politics, but we had many heroes in common, notably Winston Churchill. Our most serious

---
[1] In what follows, Armstrong's two major writings dealing with tropes in the theory of universals will be abbreviated as: *NR* for *Nominalism and Realism* (*Universals and Scientific Realism Volume 1*); and *U* for *Universals: an Opinionated Introduction*.

Peter Simons: Trinity College Dublin, email: psimons@tcd.ie disagreement outside philosophy was over Richard III. While not a full-blooded Ricardian, I had sympathies with the last Plantagenet, but David considered him an out-and-out villain and murderer, and the Tudor usurpation fully justified (we both agreed though that Henry Tudor's granddaughter Elizabeth was England's greatest monarch). It would have been interesting to have had David's reaction to the discovery of Richard's remains in Leicester and their subsequent dignified reinterment. Anyway, on to philosophy.

1 Universals and tropes

Universals and Scientific Realism appeared in two volumes in 1978. The first volume, *Nominalism and Realism*, is a thoroughgoing critique of all forms of nominalism. The second, *A Theory of Universals*, is Armstrong's constructive alternative: an Aristotelian immanent realism of universals. Unlike nearly all accounts of universals however, Armstrong's is *a posteriori*, in the sense not that arguments for it are *a posteriori* (*NR* xv), but that it is an *a posteriori*, empirical, largely scientific matter to find out which universals exist (*U* 87). That spirit of the *a posteriori* is one which I endorse throughout. Neither language nor thought dictates to us what exists, and that goes for properties and relations as well as anything else. Language and the world have to be in enough harmony for us to be able to speak truly about the latter, but this harmony is going to be extremely imperfect, so that linguistic classification and ontological classification will be considerably skew to one another, to an extent that even Armstrong did not accept. I once asked him whether it concerned him that his fourfold classification of entities into things, properties, relations and states of affairs was not suspiciously parallel to the classification of expressions into names, one- and many-place predicates, and sentences. He was brusquely unamused.

Otherwise however, we are in very close agreement. Just to list some salient metaphysical points on which I see eye to eye with Armstrong, they are: naturalism, mind-body monism, the importance of the theories of universals, the relative impotence of linguistic considerations, and the role of truth-makers both in accounting for truth and in limning ontological commitments.

In 1989 Armstrong published *Universals: An Opinionated Introduction*, a more popular but also updated account of his views, negative and positive. Like *NR*, this contained criticisms of all extant kinds of nominalism. In the meantime however, trope nominalism, of which I was by now a firm advocate,[2] had emerged as a seri-

---

[2] Simons 1982; Mulligan, Simons and Smith 1984.

ous challenger to Armstrong's immanent realism even in his own terms, so that he was prepared to accept both in print and in conversation that trope nominalism was the second best account after his own. I will argue that it is better than that.

In *NR*, Armstrong had not yet adopted the current term *trope* for properties and relations understood as particulars, but I shall employ it throughout, as he then did in *U*. Tropes have had many names, but I shall stick with this one, coined for this purpose by D. C. Williams following a perhaps ironic suggestion by George Santayana.

Tropes are particular, localized items that constitute the basis for the many ways in which concrete, substantial individuals may be like or unlike one another. Tropes have a much "thinner" nature than concrete particulars or substances. A snooker ball for example is characterized in many ways: it has such and such a mass, volume, shape, material constitution, colour, and at any time a colour, location, velocity with respect to its surroundings, magnetic and electric charge, and certain distances from other things. It, the snooker ball, is what we may call a substance: its mass, colour etc. are tropes. Suppose it is spherical, red, made of phenolic resin, has a diameter of 5.25 cm and a mass of 142 g, and is currently at 18° C. Assume for now that each of these adjectival phrases picks out a genuine existent characteristic. Whereas a realist like Armstrong would say that all these characteristics are universal, so shareable and multiply instantiable without detriment to the universals' identities, a trope nominalist says that each of them is peculiar to the ball. Another ball, even if it is exactly like this one in shape, colour, constitution, size and temperature, has its own tropes of shape, colour etc. If the ball changes in any of these respects, whereas Armstrong would say the ball comes to instantiate or exemplify different universals from the same family, a trope nominalist will say that one or more of its tropes ceases to exist in favour of another from the same family.

2 Tropes and substances

Tropes characterize substances (independent individuals) and help to make them how they are. There are two types of accounts as to how tropes relate to their substances, bundle theories and substance-attribute theories. Armstrong prefers the latter (*U* 114 ff.) I prefer the former, because it eliminates one category and one problematic formal relation of inherence or attribution. However, the most common kind of bundle theory, saying that a substance is nothing more than a bundle of tropes, is due to Williams and relies on a relation of *compresence* among tropes to tie the bundle together. If compresence is merely spatiotemporal togetherness, then I think this is wrong, because there is no guarantee that tropes have to be spatiotemporally together to be parts of one and the same substance. In quantum theory, what arrives at a telescope or photographic receptor as a single photon may earlier have been spread as wide as a galaxy, and it has been shown with some plausibility that a trope of a particle such as its magnetic moment may take a different route through experimental apparatus than that taken by other tropes such as the particle's mass.[3] For that reason, and also because in many cases it is important to distinguish a substance's essential characteristics, without which it would not be the individual it is, from its accidental ones, which it can lose and change, I prefer a double-layered account of bundles, held together by formal relations of existential dependence: a tight inner bundle or nucleus of tropes that are mutually dependent and supportive, and a looser outer layer of dispensable tropes that are one-sidedly dependent on the nucleus.[4]

On top of this nuclear bundle account of the simplest individuals one needs to recognise that very many other substantial individuals are mereologically complex, so that some of their characteristics such as mass are (to a first approximation) the resultants of those of their parts, and others, such as volume, are Gestalt tropes adhering to the complex whole. How this detail plays out in a given case depends very much on what science can tell us, which is a very Armstrongian position.

3 Armstrong's objections to trope nominalism

In *NR* Armstrong gives several reasons as to why trope nominalism fails. One is that it fails adequately to account for the resemblance of like tropes, such as the exactly resembling masses of two electrons, or colours and diameters of two snooker balls. Stout had claimed that exactly resembling tropes belong together in a class held together as a "distributive unity", a notion which Stout considers fundamental and unanalysable. Armstrong quite rightly rejects this as "a restatement of Stout's problem rather than a solution of it." (*NR* 84). It is not clear what such a unity consists in. Clearly an arbitrary class of tropes taken from here and there does not constitute a resemblance class, so we need an account of when a class is a distributive unity, and this Stout does not have.

---

[3] Denkmayr et al. 2014.
[4] Simons 1994.

The alternative view, due to Williams,[5] and endorsed by Keith Campbell,[6] is that it is the resemblance of the tropes one to another that engenders the unity of a class of resembling tropes, and not the other way round. Against this resemblance account, Armstrong has three points. The first is that the nature of resemblance depends on the nature of the objects that resemble one another, and not the other way around. The second is that resemblance leads to a vicious infinite regress, by an argument made famous by Russell, but anticipated clearly by Husserl and unclearly before him by Mill (cf. *NR* 54 n.) It is that if like tropes are as they are because they resemble one another, then these resemblances are themselves tropes, they must resemble one another, and so on ad infinitum. The third argument is that it is perfectly possible for a single trope to be of its own sort without there being any other that it resembles, and that the proposed remedy of appealing thomistically to resembling possible but non-actual tropes "is a truly desperate one." (*NR* 85).

Finally, Armstrong claims that while it is clear that one and the same particular cannot instantiate a property more than once, the nominalist needs an *ad hoc* principle to block two exactly resembling tropes from inhering in the same substance (*NR* 86).

In *U*, Armstrong adds a new objection to trope bundle theories. It is that tropes and by extension trope bundles "are not really suited to be the substances of the world" (*U* 114). A mass trope, for example, is very insubstantial, and incapable of independent existence (*U* 115).

By the end of *U*, Armstrong is ready to admit that, in a great many respects, "tropes can fill in for universals." (*U* 122). Where they fall behind, in his view, is that certain formal properties of resemblance (exact and inexact) are more satisfactorily explained by identity than by resemblance, and so by universals rather than tropes. Let us write '$a \approx_p b$' for 'a resembles b to degree p', where p is a number in the range [0,1], and the case $p = 1$ is exact resemblance. The formal properties are symmetry and (what I here call) quasi-transitivity:

SYM   If $a \approx_p b$, then $b \approx_p a$
QTRANS   If $a \approx_p b$ and $b \approx_1 c$, then $a \approx_p c$

It is Armstrong's contention that identity of properties accounts for both these more adequately than anything directly to do with resemblance (*U* 103), since trope resemblance nominalism must treat them as a "mere metaphysical coinci-

---

5 Williams 1953.
6 Campbell 1990.

dence between the properties of resemblance and the properties of identity." (*U* 137).

All of Armstrong's objections can be satisfactorily answered.

4 Answers to the objections

Let us be clear that not all versions of trope nominalism can answer Armstrong's objections easily. Stout's natural class/distributive unity account cannot. Nor does the fact that Armstrong's objections can be turned alleviate all pressures on trope nominalism. There are other matters of concern, which we shall mention later. But let us first look at the objections to trope resemblance as an account of what makes tropes as they are.

The first point is that tropes resemble one another because of their natures, they do not have natures because of what they resemble. This is right, but it does not tell against trope nominalism, because a trope is a "thin" particular: it is all nature. Substantial individuals resemble one another in manifold ways, and it is their inherent tropes that account for this. There is nothing more to a trope's being of this or that kind than that it simply exists. Being the trope it is, it cannot fail to be of that kind. A sphericality trope could not be a cubicity trope, a mass trope of 5 g could not be one of 10 g: they would just be other tropes, not the same ones with different properties. Hence, if two given tropes exist, their degree of resemblance, wherever it is between 0 (total non-resemblance) and 1 (exact resemblance), supervenes on them. Speaking in truth-maker terms, the two tropes are the joint truth-maker for their specific degree of resemblance. So if a and b are both tropes, then from the fact that both exist, it follows of necessity that they resemble one another to a certain degree, $a \approx_p b$ for a certain p. From this it follows automatically that trope resemblance of any degree is symmetric, for the truth-maker a and b just are the same two things as b and a.

Quasi-transitivity cannot be accounted for this simple way. But it is still highly motivated by the fact that it is *resemblance* that we are considering. Suppose a, b and c are three tropes, that a resembles b to degree p and that b exactly resembles c. The first fact follows from the existence of a and b, the second from the existence of b and c. Given that all of a, b and c exist, naturally both of a and c exist. Could a and c resemble one another to a different degree than a and b do? It appears impossible, because exact resemblance is precisely what guarantees that in regard to resemblance to third objects, b and c are indiscernible. Another way to put it is to say that quasi-transitivity is analytically contained in the notion of exact resemblance: it is part of what the concept of exact resemblance is there for.

This is, I think, hardly an *ad hoc* matter. Its relationship to identity will come up below.

The Mill-Husserl-Russell infinite regress argument against trope resemblance is now easily answered. If the mere existence of the terms of a case of exact resemblance suffices for the truth that they resemble one another exactly, then there is no need to invoke an additional entity, a relational trope of resemblance, to account for the resemblance. The resemblance comes automatically, for free, with the terms. Without tropes of resemblance, the regress cannot get started. And that is the best way to counter regresses. Another way to put the point is to say that the relation of resemblance (of whatever degree) between two tropes is an internal one. Personally I don't like the terminology of internal relations, since it tends to suggest that there are these items, internal relations. Since truth-making does not require such additional items, I prefer to say that the two tropes are internally related. For two or more things, being internally related in a certain way is the relational counterpart of having an essential property: they could not exist and not be so. The resemblances among more substantial individuals are not generally internal, since these turn on which tropes the substances have, some of which they may have contingently.

The point that a trope may be unique of its kind is well taken, but does not harm our view that resemblance supervenes on tropes: indeed, it is to be expected and welcomed.

Is trope nominalism at a disadvantage vis-à-vis realism in regard to how often a trope of a certain kind may be found in a substance? I think not, for two somewhat opposed reasons. Firstly, the idea that an individual cannot instantiate the same universal more than once appears to be no less questionable than that it might hold two or more tropes of one kind. The two intuitions appear to be on an even footing. It is only if tropes are thought of as little thin mini-substances that the idea of their clumping together against the regulations in deviant bundles is at all plausible. But we take tropes not to be arbitrarily recombinable "junior substances" but as dependent moments, whose freedom to wander from one bundle to another is nil and whose freedom to associate is also essentially constrained. Secondly however, the idea of two or more tropes of the same kind cohabiting in the same volume is not so far-fetched as it might appear. The two electrons in an unattached helium atom are not neatly separated into two bundles of tropes, but co-occupy the same region. So there are two mass m_e tropes and two charge $-e$ tropes, in the same place at the same time. True, there are two directionally opposed spin tropes, but no fact of the matter as to which of two electrons has which spin. Electrons are fermions, so two spatially coincident ones must diverge in at least one quantum respect by the Pauli exclusion principle. More extremely, bosons such as photons may occupy the same place at the same time and moreover be in indiscernible quantum states. This happens in lasers. So multiply collocated tropes of a type appear to be commonplace according to fundamental physics. If more than one trope of a kind is not allowed in one bundle, this would entail that such coincident particles are not genuinely distinct individual substances, a view taken by Schrödinger.[7]

Since we do not consider tropes to be substances, or substance-like in general, there is still the question as to how they can clump together so successfully as to constitute regular substances. Here I think it is more a question of getting used to the thought that they do rather than finding any deep metaphysical mystery. The basic constituents of the world are very simple in nature, and thus contain or comprise only a few tropes. Larger individuals are composed of these as parts, and what with their manifold interactions and spatiotemporal relationships, large numbers of these in proximity can plausibly manifest the familiar properties of bulky, substantial things.

5 Some advantages of trope nominalism

Taken together, these responses to Armstrong's objections at least allow the trope nominalist to breathe more easily. Trope nominalism of the sort outlined also has some genuine advantages over Armstrong's realism about universals. The principle one is ontological parsimony. All entities are particulars: the categorial division between universals and particulars is avoided. Nor is there a problem of individuation, or particularization, since tropes are particulars from the start. This feature appealed to medieval nominalists such as Ockham. Further, whereas Armstrong needs a third category, that of states of affairs, with which to bind universals and particulars together and to provide the truth-makers for elementary predications, trope nominalists require no such things.[8] The truth-making role for a great many contingent truths about particular things can safely be assigned to the tropes themselves:[9] by merely existing, this mass trope of this particle makes it true that the particle has such and such a mass. Since the mass trope cannot exist unless the particle itself exists, of which the trope is a part or in which it inheres, the particle's existence is automatically guaranteed by that of the trope, and the trope's being of this or that kind is given with its mere existence.

---

7 Schrödinger 1950.
8 Simons 2009.
9 Mulligan, Smith and Simons 1984.

The only remaining question is the nature of the relationship between trope and substance, what the medievals called inherence. Here trope nominalism also shows its mettle. A potentially looming regress that it adroitly avoids is the Bradley Regress of instantiation. Unlike realist theories, which must account for the relationship between a universal and the particulars that instantiate it, and tend to tie themselves in knots about it, with foundation or dependence replacing compresence as the glue binding trope bundles we automatically obtain inherence as another case of internal relatedness, in this case one of unilateral or multilateral dependence. This, like resemblance, is a "wireless" relationship, coming for free with the terms as a concomitant of their existence. Like resemblance, it nips the threatening regress in the bud before it can even get started.

Armstrong lays considerable store on explaining resemblance through identity, whether total or partial identity. Some of the aspects of affinity are neutral between a nominalist and a realist point of view. It is very natural to think of identity as a limiting case of resemblance: things resembling one another so closely as to be indiscernible in any respect are identical. This can be encapsulated in the two directions of Leibniz's Law:

$$a = b \leftrightarrow \forall F(Fa \leftrightarrow Fb)$$

and may even be construed as a definition of identity. Varieties of resemblance, including the exact resemblance of tropes of a kind, can be viewed as variations on the idea of limited or restricted indiscernibility, with the schema:

$$\forall F(\Phi(F) \rightarrow (Fa \leftrightarrow Fb))$$

If the antecedent condition Φ is relaxed so it becomes tautological, we regain identity. This explanation of the relationship is neutral as between nominalism and realism. A more troubled and troubling aspect of Armstrong's affection for partial identity as explicating resemblance is his later adoption in *Truth and Truthmaking* of Baxterism, the idea that the instantiation of universals by particulars consists in partial identity of them. But whereas Baxter is a Scotist about partial identity, allowing it to be contingent, Armstrong prefers a Kripkean necessity of partial identity, which threatens to undercut much of the contingency that appears to pervade the world.[10] Since Kripke is right about identity, and the world is replete with contingent things and events, it is better to treat identity as the limiting case of resemblance rather than resemblance as partial identity. Trope nominalists have to do this from the start.

---

10 Simons 2005.

6 Remaining problems for the trope nominalist

It would be disingenuous to claim that trope nominalism is clear of all difficulties simply because it can weather Armstrong's criticisms. *No* decently worked out position in metaphysics is clear of all difficulties, because metaphysics, when done properly, is fundamental, systematic, coherent,[11] and therefore hard. The problem, as Armstrong well recognised, is not one of technical complexity or the need for elaborate formal tools, as in fundamental physics, but of delicately balancing a wealth of competing considerations which are crucially important yet relatively removed from the familiarity of the everyday or the testability of scientific hypotheses. So what problems does trope nominalism still face, and where might Armstrong's realism have the advantage? I shall mention three areas of difficulty: relations, space and time, and laws of nature.

What made it true that the Titanic collided with a certain iceberg on the night of 14 April 1912? Armstrong would say it is a state of affairs. I would say on the contrary that it is the event of their colliding, that specific dated and located occurrence. A collision event requires at least two bodies as participants: it takes two to collide. So a collision depends for its existence (occurrence) on there being two bodies that collide. It thus passes the test for being a trope, but a relational one, that is to say, one dependent on two or more distinct things. This is a good case for there being some tropes that are relational. But relational particulars, unlike this case, and unlike non-relational tropes, cannot in general sensibly be assigned a location, and so their particularity and causal efficacy, hallmarks of the real for a nominalist, are debatable. One prominent trope theorist, Keith Campbell, therefore denies that there are relational tropes, while at the same time upholding the non-reducibility of relational truths.[12] I have some sympathy with this view.[13] Here is not the place to attempt to resolve the issue. Let me simply mention that one kind of relations, namely spatiotemporal ones (more below) was cited by Russell as evidence in favour of Platonic realism about universals. Given that Edinburgh is north of London, where is the *north of* bit? Russell's plausible answer is "nowhere and nowhen".[14] So if there are relational universals, some of them at least appear

---

[11] Cf. Whitehead 1978, 3: "'Coherence' [...] means that the fundamental ideas, in terms of which the scheme is developed, presuppose each other so that in isolation they are meaningless." This is perhaps the hardest desideratum to fulfil.
[12] Campbell 1990, ch. 5.
[13] Simons 2010.
[14] Russell 1911, 56.

not to be immanent in their terms in the way that non-relational universals are, and that is a prima facie problem for an Aristotelian realist of Armstrong's stripe.

Since the time of Leibniz's and Clarke's famous exchange, it has been debated whether space and time (or their unitary descendant, spacetime) are substantial or relational, Clarke following Newton and taking the former point of view, Leibniz the latter. It would not be anti-nominalist to suppose that spacetime were a single large and substantial particular, the ultimate bearer of all tropes.[15] Whether that is correct or not is another matter.

My own lightly held preference has always been for Leibniz and relationism. At one time it seemed that the possible existence of spatiotemporal vacua would be a counterargument, but in the light of quantum field theory it now seems that vacua after all do not exist: all parts of spacetime are host to *something*, most likely processes rather than enduring continuants. But relationism makes spatiotemporal relations among the occupants the primary feature, and as we have seen, for a tropist relations are not straightforward. Suppose the mug and pen on my desk are currently 10 cm apart at their closest points. What makes this true? A realist about universals would say the closest parts instantiate the relation *being 10 cm from*, as do untold many other pairs of things, so the state of affairs of the mug's being 10 cm from the pen is the unproblematic truth-maker. The status of the relation is similar to that of other universals. A trope nominalist on the other hand is put in the awkward position of considering whether there is a relational trope of the kind *being 10 cm from*. Unlike non-relational tropes, these cannot plausibly be located with their bearers. They are like Russell's *north of*, and even more embarrassingly than for an immanent realist, nominalists by their profession generally expect everything to have a spatiotemporal location, which in this case would seem absurd.

There is some relief available. Events, processes and other temporally extended items differ from enduring continuants like bodies as regards their location. Where a body is at a given time is a contingent matter. The mug is 10 cm from the pen, but they need not have been. By contrast, events etc. appear to occur essentially where and when they do. The *Titanic* might have collided with another iceberg, or with the same one at a different place or time, but the actual collision could not have taken place anywhere or anywhen else than where and when it in fact did. If this is right, then the spatiotemporal relationships between pairs of events are internal. Given the occurrence of the events, how they are related spatiotemporally comes automatically for free, and does not call out for a species of spatiotemporal trope truth-makers. That does not absolve us of the

---

15 Campbell 1990, ch. 6 advocates this.

responsibility to explain how continuants happen to be where they are when, but it shifts the problem to that of explaining how continuants are ontologically related to the events and processes that (in a process ontology) constitute them.[16] If we adopt a process ontology, as I think we should anyway, then the framework of space and time and the manifold relationships in them can be incorporated in a trope nominalist picture. But while this avoids the need for relational tropes as special truth-makers for facts of distance, angle, curvature etc., it does not tell us what such relationships *consist in*, indeed it precisely ducks that question, and I confess that for now I am stumped on how to proceed in addressing it.

In *What is a Law of Nature?* Armstrong proposed that laws consist in relationships of necessitation and probabilization among universals. This account is obviously unavailable to a trope nominalist, so it counts as an advantage of Armstrong's view that he at least has a positive theory of laws of nature, whether or not it is correct. A trope nominalist has to look elsewhere. One obvious solution would be to resort to a Humean regularity account of laws, but this has well known problems. Even if the laws of nature are not immutable, but can vary with conditions,[17] the fact remains that in the universe we can observe, things behave with extreme regularity, and that cries out for an explanation, one which the trope nominalist is at a disadvantage to give.

7 Conclusion

Trope nominalism, of the sparse, *a posteriori* type, combined with a dependence-based bundle theory of more substantial particulars, a process ontology, a non-maximalist truth-maker theory (not here discussed),[18] and a parsimonious account of internal relatedness, is a serious contender for a comprehensive metaphysics to rival Armstrong's immanent realism. It survives his criticisms, it has some advantages and a few disadvantages with respect to his position. It needs more work, but then so does any metaphysics. The ultimate aspiration of the metaphysician, a comprehensive and coherent system, is one towards which David Armstrong worked all his life, and unlike most metaphysicians, he lived to achieve it.

---

16 See Simons 2000a.
17 As advocated in Unger and Smolin 2015.
18 Cf. Simons 2000b, 2005.

Bibliography

Armstrong, D. M. 1983. *What is a Law of Nature?* Cambridge: Cambridge University Press.
Armstrong, D. M. 1989. *Universals. An Opinionated Introduction*. Boulder: Westview. (*U*)
Armstrong, D. M. 2004. *Truth and Truthmakers*. Cambridge: Cambridge University Press.
Campbell, K. 1990. *Abstract Particulars*. Oxford: Blackwell.
Denkmayr, T. et al. 2014. "Observation of a quantum Cheshire Cat in a matter-wave interferometer experiment". *Nature Communications*, 5, 4492.
Mulligan, K., Simons, P. M., Smith, B. 1984. "Truth-Makers". *Philosophy and Phenomenological Research*, 44, 287–322.
Russell, B. 1912. *The Problems of Philosophy*. London: Home University Library.
Schrödinger, E. 1950. "What is an Elementary Particle?". *Endeavour*, 9, 109–16.
Simons, P. M. 1982. "Moments as Truth-Makers". In W. Leinfellner and J. C. Schank (eds.), *Language and Ontology*. Vienna: Hölder-Pichler-Tempsky, 1982, 159–61.
Simons, P. M. 1994. "Particulars in Particular Clothing. Three Trope Theories of Substance". *Philosophy and Phenomenological Research*, 54, 553–76.
Simons, P. M. 2000a. "Continuants and Occurrents". *The Aristotelian Society*, Supplementary Volume LXXIV, 78–101.
Simons, P. M. 2000b. "Truth-Maker Optimalism". *Logique et analyse*, 169–170, 17–41.
Simons, P. M. 2005. "Negatives, Numbers and Necessity: Some Worries about Armstrong's Version of Truthmaking". *Australasian Journal of Philosophy*, 83, 253–261.
Simons, P. M. 2009. "Why there are no States of Affairs". In M. E. Reicher, ed., *States of Affairs*. Frankfurt/Main: Ontos, 111–28.
Simons, P. M. 2010. "Relations and Truthmaking". *The Aristotelian Society*, Supplementary Volume LXXXXIV, 199–213.
Unger, R. M., Smolin, L. 2015. *The Singular Universe and the Reality of Time*. Cambridge: Cambridge University Press.
Whitehead, A. N. 1978. *Process and Reality*. Corrected edition. New York: The Free Press.
Williams, D. C. 1953. "The Elements of Being". *Review of Metaphysics*, 7, 3–18, 171–92.

Anna-Sofia Maurin
Tropes: For and Against

1 Introduction

Trope theory is the view that the world consists (wholly or partly) of particular qualities, or tropes. This admittedly thin core assumption leaves plenty of room for variation. Still, most trope theorists agree that their theory is best developed as a one-category theory according to which *there is nothing but tropes*. Most hold that 'sameness of property' should be explained in terms of resembling tropes. And most hold that concrete particulars are made up from tropes in compresence (for an overview, including an introduction to some alternative versions of the view, cf. Maurin, 2014). D. M. Armstrong disagrees. He thinks the world is a world of immanent universals, the thin particulars in which those universals are instantiated, and the states of affairs that – thereby – exist.[1] He holds that 'sameness of property' should be explained in terms of numerically identical universals or – if the resembling things are the universals themselves – in terms of partially identical universals. And he believes that concrete particulars are made up from 'thin particulars' in which a (sufficient) number of universals are instantiated. In spite of their disagreements, proponents of tropes owe Armstrong a debt of gratitude. First, for being one of the theory's earliest, most serious, and – not least – most prominent, critics (Armstrong for the first time considers, and rejects, what he then labels 'particularism', in his 1978a). But also for being one of the theory's most ardent champions. Already in his 1989a, Armstrong regrets his 1978-rejection of the view. Equivalence classes of exactly resembling tropes, he now admits, for most purposes "serve as an excellent substitute for universals" (Armstrong 1989a, 122). But then why isn't Armstrong a trope theorist?

In this paper, Armstrong's main reasons for preferring universals to tropes are critically scrutinized. All of them, it is maintained, fail to convince. If this argument is accepted, Armstrong seems to have no – or, at least, no good – reason for not accepting the existence of tropes. But, then, ought Armstrong to have been a trope theorist? If the 'best' version of the trope view turns out to be ontologically

---

[1] That a state of affairs exists so-to-speak *on top of* the thin particular and universal that makes it up is something Armstrong *must* (and does) accept (but cf. his 1978a, 80). For a critical discussion cf. Maurin 2015.

Anna-Sofia Maurin: University of Gothenburg, email: anna-sofia.maurin@gu.se

more parsimonious than Armstrong's own theory of universals, then, yes. That there is a version of the trope view, a version that is not (or, at least not seriously) considered by Armstrong, that is ontologically more parsimonious than the universals view, is argued in the next section. Why none of Armstrong's reasons for preferring universals to tropes manage to convince, is explained in the section after that. First, however, and in order to avoid a common misunderstanding of the trope view a few more words about the core-difference between tropes and universals, and about how this difference matters (as well as does not matter) to what – and how well – these theories explain.

2 Tropes and the one over many

The core-difference between tropes and universals, first, is rather easily stated: tropes are particular, universals are universal. Among other things, this means that, if a and b are both F, and F is a universal, then a and b *literally* have something in common: they both exemplify (numerically the same) F. Not so if properties are tropes. For then, if a and b are both F, this is because a exemplifies trope f_1 and b exemplifies trope f_2, where $f_1 \neq f_2$. What, then, makes a and b the same? If properties are tropes, a and b are the same if a exemplifies f_1, b exemplifies f_2, and f_1 (exactly) resembles f_2.[2]

According to Armstrong, the most important job for any theory of properties is precisely the one described above: that of explaining (or 'grounding') the fact that things which are distinct, nevertheless appear to be the same ('in some respect'). This is the 'problem of the One over Many', and solving it, Armstrong tells us, is a 'Moorean' task.[3] This problem is also a reason not to be a 'classic' nominalist. For, if you are, then because you refuse to accept the existence of a separate category of properties you have no option but to explain the sameness of a and b with reference to something external to them (like their being members of a certain class or falling under a certain predicate).[4] But this is problematic. First, because it turns

---

[2] Not all trope theorists are resemblance theorists, even though most are (cf. e.g., Stout (1921, 1923)).

[3] A problem is 'Moorean', according to Armstrong, if a theory, to be acceptable, must be able to substantially solve it (Armstrong 1980, cf. also Devitt 1980). Armstrong does not think that a solution which takes the contested phenomenon as primitive should count as substantial. This is why he considers the Quinean nominalist an 'ostrich' hiding his head in the sand, refusing to answer a compulsory question. This view is criticized in e.g., Lewis 1983.

[4] A possible exception to this, at least according to Armstrong (1989b), is classic resemblance nominalism. Cf. fn 6 below.

explanation on its head: intuitively, *a* and *b* are both F not because they belong to the same class or fall under the same predicate. They belong to the same class or fall under the same predicate because they are both F.[5] More importantly, it is problematic because what the nominalist is in effect doing, is accounting for the sharing of one type – 'being F' – in terms of the sharing of another – 'belonging to the same class', 'falling under the same predicate', etc. But the nominalist, being a nominalist, cannot accept the existence of any types. Therefore, these new types must in turn be explained (away). And if this (new) explanation is couched in nominalist terms (which it certainly ought to be), one type is, once again, explained with reference to another. Which means that we have yet another type that must be explained (away). And so on, *ad infinitum*. At no point is the problem of the One over Many substantially solved. And this, Armstrong concludes, is unacceptable (for Armstrong's classic attack on nominalism cf. his 1978a, pt. 2).

Properties help. For if there are properties, Armstrong argues, then that *a* and *b* are both F can be explained with reference to the properties they have, which means that explanation is turned right back on its feet. How about the threat of vicious infinite regress? If properties are universals that threat is averted. For if properties are *identical* in their distinct instances their being exemplified is enough to ensure that those instances are 'the same'. Matters are not quite as straightforward if properties are tropes. For then, as we have seen, to explain the sameness of *a* and *b*, it is *not* enough to point to the property-tropes they exemplify. We also need an account of what makes those tropes the same. Just like the classic nominalist, the trope theorist must provide the requisite account in terms that respect the basic tenets of her theory. To avoid positing universals, therefore, she must take whatever it is that makes two tropes 'the same' – their resemblance, on the standard view – as yet another trope. But then, for each pair of resembling tropes, there will be one resemblance-trope. Provided that there is more than one pair of resembling tropes (which seems likely), there will be more than one resemblance-trope. But those resemblance-tropes are also the same (they are all resemblances). The trope theorist must now explain *their* sameness, and she must do so in trope-theoretical terms. That is, she must explain their sameness in terms of their resemblance. But this saddles her with yet another set of resemblance-tropes. Which resemble. And so on, *ad infinitum* (Armstrong 1978a, 85, cf. also Maurin 2013).

Is this a vicious regress? According to a number of trope theory's critics, it certainly is. However, after briefly toying with the idea that it is in his 1978, Armstrong post-1978 does not agree. Why not? The short answer is: because of the nature of resemblance. Resemblance, it is generally agreed, is an *internal relation*. As such,

---

5 A so-called 'Euthypro dilemma' (from Plato's *Euthypro*, 10a). Cf. Armstrong 2004, 40.

it is grounded in the nature of that which it relates. If properties are universals, the nature which grounds resemblance is numerically the same in its different instances. If properties are tropes, it is not. However, post-1978, Armstrong does not think this difference matters to the workings of resemblance. Just like if properties are universal, if properties are tropes, resemblance "flows from the nature of the resembling things" (Armstrong 1989a, 44).[6] But if resemblance "flows from the natures of the resembling things", the threat of *vicious* infinite regress disappears. For now "[t]he truth-maker, the ontological ground, that in the world which makes it true that the tie [i.e., resemblance] holds, is simply the resembling things" (Armstrong 1989a, 56). The resemblances in which the resembling things stand will still resemble each other; and those resemblances will resemble each other, and so on *ad infinitum*. But as the existence of these resemblances is entailed *given* the tropes posited in the regress' first step, their existence cannot prevent those tropes from solving the problem of the One over Many. Contrary to what sometimes appears to be the popular opinion, therefore, Armstrong *doesn't* think that the resemblance regress is a reason to prefer universals to tropes (although, as we shall see, he *does* think that resemblance may provide the universal realist with another kind of reason to that effect).

3 Armstrong on what is the 'best' version of the trope view

Apart from their shared acceptance of the existence of (fundamental) tropes, trope theorists differ – sometimes widely – among themselves: about what else there is besides tropes (if anything), about how tropes can make true propositions ostensibly about universals, and about how tropes can make true propositions ostensibly about concrete particulars. And so on.[7] The following is however a popu-

---

6 According to Armstrong (1989b), this move can be used to save also the classic resemblance nominalist (RN) from the usual nominalist critique. However, if the trope view has resemblance 'flow' from the tropes had by the resembling objects, RN must "congeal the particular properties into a single grand (but still particular) property within which no differentiation can be made" (Armstrong 1989a, 45). Somewhat surprisingly, Armstrong doesn't think accepting particularized natures had by the resembling things contradicts the basic tenets of RN. He does admit that they are 'a somewhat blunt instrument', however, and this is taken as a point in favor of the trope-view (*ibid.*)

7 Some have even argued that trope theorists differ in exactly what *kind* of thing they take the trope to be. For recent statements of this view, cf. Loux 2015, and Garcia 2015.

lar – probably *the* most popular – version of the theory, and so deserves to be called 'the standard version': (i) there are only tropes; (ii) propositions ostensibly about universals are made true by tropes *in resemblance*; (iii) propositions ostensibly about concrete particulars are made true by tropes *in compresence*.[8] But Armstrong's preferred version of the theory, the one with which he compares his own immanent realism, is *not* this one. Although, as we have seen, he accepts (ii), Armstrong rejects (iii). Propositions ostensibly about concrete particulars, he holds, are on the best version of the theory made true not by bundles of compresent tropes, but by tropes exemplified by (what he calls) thin particulars.[9] This means that, on Armstrong's preferred version of the theory, not just tropes but also thin particulars are taken as fundamental, which means that (i) is abandoned. Worse, because Armstrong rejects an assumption common among the minority of trope theorists who, like him, prefer a substance-attribute account of concrete particulars, namely *that tropes are non-transferable*, he also thinks that the trope theorist must accept the existence of a *third* kind of thing: states of affairs. All in all this makes Armstrong's preferred version of the theory ontologically speaking much more expensive than the standard version.

Armstrong lists three reasons for why one ought to prefer a version of the trope view that rejects (iii) (and, hence, (i)): (1) tropes are ways things are (cf. Armstrong, 1989a, 115f and 1997, 25); (2) tropes are not substances (cf. Armstrong 1989a, 114f.), and; (3) objects do not have their properties of necessity (cf. Armstrong 2004, 46). Take (1) first. This is a reason for rejecting (iii) because, if tropes are ways, then they are essentially of some object, they are 'characterizers', and so stand in need of something to characterize. This can straightforwardly be made sense of if concrete particulars are understood along substance-attribute lines: tropes are of, and so characterize, the thin particulars that exemplify them. Not so if concrete particulars are bundles of tropes. For then what is there for those tropes to be of? (1), moreover, relates rather intimately to (2). For, if tropes are essentially of some object, then whatever else this means, it certainly seems to entail that they are essentially dependent entities. But, Armstrong believes, on the bundle view, tropes must be taken as some kind of 'junior substance': if tropes are all there is to the concrete particular, then there is literally nothing (no other kind of thing) for those tropes to depend upon (cf. e.g., Armstrong 1989a, 115). And (3), finally, is a reason to be a substance-attribute theorist because, to a bundle theorist "[p]redication of the member [of a bundle] is a mere matter of extracting the trope or the universal

---

8 Classic proponents of the 'standard' view are Williams 1953, and Campbell 1990.
9 This does not stop Armstrong from acknowledging the many virtues that a bundle-of-tropes account has in comparison to a bundle-of-universals account (cf. e.g., Armstrong 1989a, 114).

from the bundle" (Armstrong 2004, 46).[10] If the concrete particular is the bundle of its tropes, therefore, those tropes could not be replaced, yet that object continue to exist. This is highly unintuitive.

None of these are good reasons for rejecting (iii), however. That tropes are *ways things are*, first, is only a reason for adopting the substance-attribute view instead of the bundle view if the 'things' of which tropes are supposed to be 'ways' can be reasonably understood as *thin* particulars. But it seems highly unlikely that they can be. To describe properties as essentially *of* some object, could just as well (better, even) be understood as saying something about how properties stand to the *concrete* particular that 'has' them. And concrete particulars are *thick* not thin particulars. How about dependence? If tropes are ways, they are essentially dependent entities. But, as Armstrong points out, if there is nothing but tropes, then there is nothing – no other kind of thing – for those tropes to be dependent upon. How, then, can they be ways in the first place? This is how: by being such that they must belong to some bundle. Or, in other words, by being such that they, in order to exist, must be compresent with some other trope(s).[11] For this would mean that, even if they do not depend for their existence on the existence of an entity belonging to another *kind*, tropes nevertheless essentially depend for their existence on something. Which should be enough to accommodate the intuition that tropes – in being ways – are essentially dependent entities.[12] How about Armstrong's final point about essential predication? Although Armstrong is quite right in pointing out that if the concrete particular is identified with the bundle of its properties, then *it* could not be differently constituted, he does not seem to realize that this is a problem – if it is a problem – for *any* theory according to which the concrete particular is (at least in part) identical with its own properties. Which means that this is a problem for *both* the bundle- and the substance attribute view (provided, that is, that constitution is understood as (sufficiently like) identity).

Armstrong also thinks that, on the 'best' version of the trope view, tropes are *transferable*. Again, this is not the standard view, at least not among the minority of trope theorists who, like Armstrong, accept a substance-attribute view of

---

[10] Given Armstrong's overall methodological framework, talk of 'predication' here is best understood in non-linguistic terms. This is also how I've taken it above.

[11] This is in fact the 'standard' view among most trope-bundle theorists.

[12] Note, also, that Armstrong could hardly object that, if tropes are essentially dependent entities, then whatever bundle they make up must itself be a dependent entity – which would leave the one-category trope theorist with a universe devoid of substance. He could not, because, on his own view, states of affairs are made up from (mutually dependent) thin particulars and universals, yet states of affairs are essentially independent.

concrete particulars.[13] On the standard view (in this sense of 'standard', that is), tropes are *non*-transferable, i.e., they are such that, if they exist and (partly) constitute a concrete particular, there is no possible world in which they exist and fail to do so. From the existence of the constituents of a concrete particular, in other words, the existence of that concrete particular is entailed. Armstrong presents two reasons against the standard view. If tropes are non-transferable, first, what he believes is the best available theory of modality (which requires free recombination of all items in the actual world, cf. his 1989b) would arguably have to be given up. But even if this account of modality is rejected, Armstrong points out, a world with non-transferable tropes would be a world that is 'ineluctably fixed'. A 'rather mysterious necessity' that Armstrong thinks ought to be avoided if at all possible (Armstrong, 1989a, 118; cf. also Maurin 2010). However, if tropes are transferable, and this is explicitly admitted by Armstrong, then the trope theorist must posit, not just tropes and thin particulars, but states of affairs as well. This is why (Armstrong 1989a, 117; cf. also Armstrong 1991, 193):

> suppose that *a* has property trope F. This is either a matter of F's standing in the bundling relation to the other tropes that make up *a* (bundle version) or else is a matter of F's being an attribute of *a* (substance-attribute version). In either case, states of affairs are required. For instance, *a*'s being F entails the existence of *a* and trope F. But *a* and trope F could exist without *a*'s being F. So, [*a*+F] (the object that is the mere sum of *a* and F) is an insufficient truth-maker for *a*'s being F. States of affairs are required as part of the ontology of any trope theory.

If tropes are *non*-transferable, on the other hand, then *a* and trope F *could not* exist without *a*'s being F, thereby removing any need for states of affairs (Armstrong 1989a, 118). Non-transferable tropes have their cost. Primarily, a weirdly fixed universe. But transferable tropes, we can now see, have *theirs*. "Which poison should the boys in the backroom choose?" Armstrong asks, knowing full well which one he prefers (Armstrong, 1989a, 118). But why should the trope theorist agree? As we have seen, with non-transferability hers is the more parsimonious view. With concrete particulars understood along bundle- instead of substance-attribute lines, it is more parsimonious yet. But then, supposing that no *other* reason for preferring universals to tropes can be supplied, parsimony might be what tips the balance in trope theory's favor.

---

[13] There are also bundle theorists who think that tropes are non-transferable. One example is Simons 1994, who attributes non-transferability to the tropes in what he calls the "kernel" of the bundle.

4 Armstrong on why there are no tropes

To see if some reason (of sufficient strength) for preferring universals to tropes can be supplied, or, at least, to see if Armstrong succeeds in supplying one, we need to investigate more carefully the reasons Armstrong does provide in support of his contention. Here they are in very brief summary:[14]

> *Piling*: on the trope view, you must either accept 'piling' or you must accept an *ad hoc* principle forbidding piling. The former is unintuitive, the latter is theoretically costly.
>
> *Swapping*: on the trope view, you must either accept 'swapping' or you must accept that tropes are non-transferable. The former contradicts a fundamental Eleatic principle that Armstrong thinks we ought to accept, the latter (as we have just seen) forces us to accept that the universe is (mysteriously) fixed.
>
> *'Hochberg's Argument'*: on the trope view, if f_1 and f_2 are distinct yet exactly similar tropes, then the propositions *that f_1 and f_2 are exactly similar* and *that f_1 and f_2 are distinct*, although formally distinct, must be given the same truthmakers (namely f_1 and f_2). This is unintuitive.
>
> *Laws of Nature*: on the trope view, Armstrong's 'relational' theory of laws of nature must be rejected in favour of a regularity account. But this means that the trope theorist must accept an inferior theory of laws.
>
> *Resemblance*: on the trope view, the axioms of resemblance as well as the axioms of identity must be taken as primitive. This is theoretically uneconomical.

Armstrong doesn't think of any one of these reasons as conclusive. What he does believe, however, is that they, especially if considered 'in bulk', strongly suggest that a theory that admits universals – an *immanent realism* – is preferable to one that admits tropes (cf. e.g., Armstrong 1997, 24). I think Armstrong is mistaken. First because, as I will try to convince you next, piling, swapping, and 'Hochberg's argument' (at least as that argument is presented by Armstrong) most likely do not

---

14 One reason that I won't discuss here is what we may call Armstrong's 'argument from robustness', which claims that universals ought to be preferred to tropes because with their help, the problem of the One over Many can be solved in an unusually 'robust' manner (cf. esp. Armstrong 1997, 22; cf. also his 1978a, xiii). Armstrong throughout his writings seems to hesitate over just how seriously this argument should be taken. That accepting its spirit probably means contradicting other theses he holds is argued in Maurin 2008.

provide the universal realist with *any* reason for preferring universals to tropes. Second, because laws of nature and resemblance at best provide her with very weak such reasons. Reasons, moreover, that in order to *be* reasons, require us to make what many would consider to be highly contentious and/or theoretically costly assumptions.

5 Piling, swapping, and 'Hochberg's argument'

Why don't I think that piling, swapping, and 'Hochberg's argument' provide the universal realist with reasons to prefer her posits over those of the trope theorist? Take 'piling' first. Piling is what would be the case if one and the same particular (at one and the same time) instantiated a property more than once. Piling, Armstrong believes, is impossible (or at least 'highly improbable'), for "[t]o say that *a* is F and that *a* is F is simply to say that *a* is F" (Armstrong 1978a, 86). If properties are universals, that this is impossible follows automatically: if *a* is F and *a* is F, then, since F=F, what this boils down to is simply that *a* is F.[15] Not so if properties are tropes. To make sure that the possibility of piling is ruled out, therefore, Armstrong thinks that the trope theorist must introduce an *ad hoc* principle that forbids it. *Ad hoc* principles are bad, and we ought to prefer a theory on which they are not needed. Therefore, Armstrong concludes, we ought to prefer universals to tropes. But note that piling only forces the trope theorist to accept an *ad hoc* principle, if avoiding piling is as important as Armstrong suggests. It is however unclear that it is. It is, in other words, unclear if piling *is* an empty possibility. Suppose that piling is possible. Then, if the apple is red, this is either because it exemplifies one, or two (or three, or…) red-tropes of some determinate shade of red. Although piling tropes on top of tropes in this way certainly seems unnecessary, it does not lead to any real trouble. None of the tropes, however many, prevent us from completing any important explanatory task, or from making any necessary perceptual discriminations. *That a is F* is made true by the tropes had by *a*, whether those include one, two, or infinitely many red-tropes. But then, if nothing prevents the possibility of piling, and if this possibility makes no difference to anything we might care about, why must we accept an *ad hoc* principle that forbids it? The short answer is that we must not. But then piling is not a reason to prefer universals to tropes.

---

15 But what doesn't follow 'automatically', is that a universal realism of the substance-attribute kind can avoid the possibility of 'piles' of thin particulars (at least not without accepting some *ad hoc* principle that forbids them). Thank you Daniel Giberman for pointing this out.

Swapping, next, is (as far as I have been able to establish) first discussed by Armstrong in his 1989a. To test our intuitions, Armstrong first invites us to consider a case in which *a* has property P but lacks Q, while *b* has property Q but lacks P. In this case, he claims, it makes perfect sense to say that *a* might have had Q and not P, while *b* might have had P but not Q. But if properties are tropes, this means that the following situation (where P' and P" are exactly similar tropes) ought to make sense as well: *a* has property P', *b* has property P", but *a* might have had P" and *b* might have had P'. But, Armstrong thinks, it does not. For, he points out, as "[t]he swap lies under suspicion of changing nothing" (1989, 131–2), it contradicts a fundamental Eleatic principle according to which only what makes a causal difference to the world ought to be admitted in ontology (Armstrong 1978b, 45–6). Fundamental principles ought not to be contradicted. Therefore, swapping is a reason to prefer universals to tropes. One way around this is of course to hold that tropes are non-transferable. But, Armstrong points out (again), "this restricting of the way that possibilities are preserved under recombination is equally an ontological cost for a trope theorist" (Armstrong 1989a, 132). Against this argument a number of things can be said. First, one may object to its use of the Eleatic principle. For, even if this is a reason not to introduce anything in ontology that cannot or even that does not make a causal difference to the world, it is unclear (to say the least) if this means that one is thereby forbidden from postulating anything that, although not causally inert, could be involved in causally inert transactions. Also, is it really true that the swap makes for absolutely no difference? Some have argued that this is not true. So, for instance, does Labossiere (1993, 262) claim that, at most, swapping makes for absolutely no detectable (or verifiable) difference. If he is right, swapping is only a reason to prefer universals to tropes, if undetectable differences turn out to be especially unpalatable. Claiming that they are, however, requires separate argument. Moreover, in order to avoid swapping, it is unclear if the trope theorist must accept non-transferability of the, to Armstrong at least, objectionable kind. After all, in order to avoid the situation sketched above, it is arguably enough if we introduce the following restriction: if (in, say, the actual world @) *a* has P' at time t, then there is no time t' (also in @), where t ≠ t', at which *b* has P' (cf. Cameron 2006, 99–100). But this is non-transferability of an utterly weak kind. One that doesn't leave us with a strangely fixed universe, and so arguably one that Armstrong ought to have no special reason to reject (cf. Maurin 2010, 317–21). Swapping, therefore, either doesn't contradict the Eleatic principle, or avoiding contradicting that principle doesn't require accepting a strangely fixed universe. Either way, swapping is no reason to prefer universals to tropes.

What he refers to as 'Hochberg's argument', finally, is first discussed by Armstrong in his 2004.[16] Here is the argument in Hochberg's words (Hochberg 2004, 39, cf. also his 2001):

> Let a basic proposition be one that is either atomic or the negation of an atomic proposition. Then consider tropes t and t* where "t is different from t*" and "t is exactly similar to t*" are both true. Assume you take either "diversity" or "identity" as primitive. Then both propositions are basic propositions. But they are logically independent. Hence they cannot have the same truth makers. Yet, for a trope theory [...] they do and must have the same truth makers. Thus the theory fails.

That t and t* make true both *t is different from t** and *t is exactly similar to t** is problematic, says Hochberg, because it means that a fundamental principle, central to the way we think about truth and meaning, must be rejected. Let's call this principle 'Hochberg's principle' (HP):

(HP) Logically independent atomic propositions must have distinct truth makers.

If HP is accepted, Hochberg's argument does *more* than merely provide us with an inconclusive reason against the trope view; it conclusively refutes it. The trope theorist, when faced with this argument, is therefore forced to take action. One option is to simply refuse to accept HP (cf. e.g., Mulligan, Simons and Smith 1984, 115). Another is to distinguish between formally and materially independent propositions, and claim that HP should only be taken to apply in case the propositions are not just formally but also materially independent (cf. e.g., Macbride 2004). None of these options will be discussed here (but cf. Hochberg 2004 and Maurin 2005). For, interestingly enough, whatever stand one takes on Hochberg's argument, the argument Armstrong calls 'Hochberg's argument' is interestingly different from the original. This is because Armstrong does not want to accept HP in general, and so does not necessarily regard a theory that contradicts it as essentially flawed. Instead he prefers to "argue simply from a case" (Armstrong 2004, 44; cf. also his 2005). The case he asks us to consider, more precisely, is basically the same as the one discussed by Hochberg. A case, that is, in which there are two simple but exactly resembling tropes (Armstrong 2004, 44):

> Given the existence of *a* and *b*, then it is a truth, a necessary truth, that <*a* is diverse from *b*>. I would say that the truthmakers are just the mereological sum of the terms: *a* + *b*. Hochberg would disagree, but this disagreement does not seem important for the purposes of this argument. Given the existence of *a* and *b*, then it is, by hypothesis, a truth that <*a* is exactly

---

[16] Cf. also Maurin 2005, which discusses a slightly different version of Hochberg's argument.

similar to *b*>. Again, I would say that the truthmakers are *a* + *b*: the very same truthmakers for the two different truths. Given that *a* and *b* are simples, this seems counter-intuitive.

'Hochberg's argument' is much harder to make sense of, and therefore also much more difficult to evaluate, than Hochberg's (original) argument. For if this is not an argument from HP, then what, exactly, is going on here? The situation "seems counter-intuitive", Armstrong tells us. But, why? Without a principled underpinning (like HP), the 'argument' amounts to nothing much at all. Or it amounts to something very much like begging the question. For if it is not because it contradicts some underlying principle that the situation is regarded as highly suspicious, then it is most likely because *normally* distinct truths are given distinct grounds. But what does *normally* mean? Universals are entities able to exist 'fully' in more than one place in space at one moment in time. Tropes are entities such that the same tropes sometimes make true formally distinct propositions. If you think that, *normally*, things do not exist in more than one place in space at one moment in time, and if you think that, *normally*, formally distinct truths have distinct truthmakers, then neither universals nor tropes behave normally. But so what? You simply cannot fault universals for being universals, just like you cannot fault tropes for being tropes. Therefore, 'Hochberg's argument' does not provide the universal realist with a (non-question-begging) reason to prefer universals to tropes.

6 Laws of nature and resemblance

The cases of laws of nature and of resemblance are a bit more complicated. Take laws first. These provide the universal realist with a reason in favor of her view, Armstrong believes, since, if properties are universals, "a very much more plausible theory of the nature of laws of nature" can be formulated (Armstrong 1997, 24; cf. also Armstrong 1983, Dretske 1977, and Tooley 1977). On this view, if it's a law of nature that *all Fs are Gs*, this is because a relation of 'contingent (or 'nomic') necessitation' (N) holds between (*universals*) F-ness and G-ness. This view, Armstrong argues, is not available to the trope theorist, at least not without significant loss of explanatory power. At best, using equivalence classes of exactly resembling tropes as a substitute for Armstrong's F-ness and G-ness, the trope theorist "can put forward a quite plausible general principle of 'like causes like'" (Armstrong 1997, 24). What she cannot do is come up with a 'suitable truthmaker' for that principle. At least, she cannot produce a truthmaker that is not a 'mere regularity'. Armstrong explains (Armstrong 1997, 237):

What, in particular, can the truthmaker for true law-statements be? The principle required is that exactly resembling tropes should bestow the very same nomic powers. But what is the truthmaker for this principle? Can it be anything more than the state of affairs that all the token state of affairs do in fact behave in accordance with this principle? And what is this but a Regularity theory of laws?

If properties are universals, on the other hand, the 'like causes like' principle *is* adequately grounded. It is grounded, moreover, in the state of affairs N(F,G). And this is why universals are preferable to tropes. Armstrong's argument from laws of nature rests on a number of substantial assumptions. First, it assumes that tropes cannot function as adequate truthmakers for the 'like causes like' principle. Truthmakers, that is, that are not 'mere' regularities. Second, it assumes that 'mere' regularities aren't acceptable truthmakers for that principle in the first place. And, third, it assumes that a theory of laws of nature of Armstrong's variety is without (serious) problems of its own. All of these assumptions may be questioned. Here we can only begin to scratch the surface.

That tropes, if considered as powers, might function as adequate truthmakers for the 'like causes like' principle, is a possibility Armstrong considers (and then rejects) in his 2004. On this view, what it is to be an F-ness trope is to be such that you (are likely to) produce something that is (characterized by) a G-ness trope.[17] But this means that, that tropes belonging to the same resemblance class give rise to tropes belonging to the same resemblance class (the infamous regularity), is now grounded in the nature of the relevant tropes. Which means that the 'like causes like' principle can be given what, to Armstrong, are acceptable truthmakers. If viable, Armstrong admits, this view therefore provides the trope theorist with a 'satisfactory substitute for the generality that laws ought to exhibit' but without having to posit the existence of universals (Armstrong 2004, 132–3).

But Armstrong rejects the powers view. Not all properties, not even all scientifically respectable properties, he argues, can be treated as powers. For if they were, "potency [would] never issue in act, but only in more potency", which would be absurd (Armstrong 2004, 139). And the alternatives – that the world contains *both* power- and categorical tropes (the mixed view), or that it contains tropes that are simultaneously powers-and-categoricals (the aspectual view) – are likewise problematic. What, on the mixed view, is a purely categorical property supposed to be? It must have some power to affect things around it causally, or it, apart from being entirely unknowable, will offend against the, according to Armstrong truly fundamental, Eleatic principle. But then whatever causal efficacy it has, it must have

---

[17] This view is in fact defended by a number of trope theorists (cf. e.g., Martin 1980, Molnar 2003, and Heil 2003).

only contingently, or categoricals collapse into powers. Which, if at least some of the properties that figure in at least some of the laws are categorical, most likely means that some laws will have to be understood in terms of 'mere' regularities after all (for details, cf. Armstrong 2004, 140). If the trope is understood along aspectual lines, on the other hand, how does its power-side relate to its categorical side? Is their connection contingent? Is it necessary? Again, both options lead to trouble (for details, again, cf. Armstrong 2004, 141). All of these problems, Armstrong concludes, are most likely unsolvable. Therefore, regarding tropes as powers is not an option.

Suppose, therefore, that tropes cannot in any straightforward way replace Armstrong's universals in Armstrong's account of laws. This is a problem, according to Armstrong, because it means that the trope theorist must make do with a theory of laws (the regularity view) with inferior explanatory value. But this is only because, as we have seen, Armstrong thinks that the 'like causes like' principle *cannot* be adequately explained by or grounded in 'mere' regularities. Why not? Explanation must come to an end somewhere. Armstrong thinks that this end should come one step *after* the observed regularities. But the regularity theorist does not agree. Who is right will depend on which basic assumptions – about explanation, about ontological parsimony, and about theoretical simplicity (etc.) – one accepts. The issue is therefore much more complicated than it may at first appear. Note also that, just like the regularity view is not without its defenders (cf. Carroll 2012 for a good overview), Armstrong's theory is not without its critics. Among other things, it has been pointed out, it is unclear if, in positing N, Armstrong does a much better job explaining than do his rivals. In Lewis' words (Lewis 1983, 366; cf. also van Fraassen, 1989):

> Whatever N may be, I cannot see how it could be absolutely impossible to have N(F,G) and F*a* without G*a*. (Unless N just is constant conjunction, or constant conjunction plus something else, in which case Armstrong's theory turns into a form of the regularity theory he rejects.) The mystery is somewhat hidden by Armstrong's terminology. He uses 'necessitates' as a name for the lawmaking universal N; and who would be surprised to hear that if F 'necessitates' G and *a* has F, then *a* must have G? But I say that N deserves the name of 'necessitation' only if, somehow, it really can enter into the requisite necessary connections. It can't enter into them just by bearing a name, any more than one can have mighty biceps just by being called 'Armstrong'.

If laws of nature provide the universal realist with any reason at all for rejecting tropes, it is therefore a rather weak, and certainly a theoretically loaded one. A reason, that is, which shouldn't keep the trope theorist up at night.

Armstrong's argument from the axioms of resemblance, finally, is the one he himself finds the most compelling. The argument is first mentioned in his 1978a.

Here it is introduced as a reason not to be a 'classic' resemblance nominalist, but Armstrong clearly considers this a problem also for the trope view. The problem, more precisely, is that, on any view according to which sameness of property is understood in terms of resemblance, all of resemblance's formal (and other) properties must be accepted as bedrock primitives. This is problematic, first, because then, what is taken as not further explicable (or even as 'inexplicable') really isn't. Rather (Armstrong, 1978a, 49; cf. also his 1989, 57 and 137; and his 1997, 23):

> [i]t is natural to derive the symmetry of resemblance from the symmetry of identity. a resembles b if and only if a and b are in some respect identical. There exists a respect, C, in which a and b are identical. The symmetry of C's identity with itself then ensures the symmetry of resemblance.

And it is problematic, second, because, since the things that resemble each other must be *self*-identical, the formal (and other) properties of identity must *likewise* be taken as bedrock. But then not one, but at least two sets of axioms will have to be accepted as 'brute' and not further explicable. The same is not true if you are a universal realist. For on this view, resemblance is explained in terms of identity, which means that only the axioms of identity need to be taken as primitive. This makes universal realism the preferable view, because "[i]n philosophy, as in science, the theory that explains by appealing to the least number of principles is to be preferred, other things being equal" (Armstrong 1997, 23). According to Armstrong, in other words, resemblance is a reason to prefer universals to tropes, because resemblance can be accounted for in an overall theoretically simpler way by the universal realist than it can be by the trope theorist. But can it really? Only if the universal realist can account for *all* cases of resemblance in terms of identity. And only if she can do this without overly complicating her theory.

First point first. As we have seen, if properties are universal, there is a perfectly natural sense in which resemblance between concrete particulars can be understood in terms of identity: a and b resemble each other if they (literally) share a property. Not all cases of resemblance are that straightforward, however. Take two perfectly determinate shades of redness that resemble each other to degree D. It seems wrong – or, at least, it doesn't seem obviously right – to explain their resemblance in terms of some 'respect' they share. Especially as it is far from clear that universals have respects in the first place. That resemblance between properties is a more difficult case than is resemblance between concrete particulars is readily admitted by Armstrong himself (in fact, he would go as far as to say that this is "one of the most difficult issues in the theory of universals" (Armstrong 1997, 47; cf. also his 1989a, 124)). Rather than account for the resemblance between properties in parallel with how he accounts for the resemblance between concrete particulars (in terms of 'identical respects'), Armstrong opts for accounting for it in terms of 'identical part', or, in other words, in terms of partial identity: (universals) F and G resemble each other if they literally share some part(s). In Armstrong's words (1989a, 106, cf. also his 1978b, 116–30):

> Being five kilos in mass involves the five-kilo thing having a part, a proper part to put it technically, that is four kilos in mass… The properties resemble because a four-kilo object is a large proportion of a five-kilo object. The bigger the part, the closer the identity, and so the closer the resemblance.

So far this sounds rather straightforward. However, the account soon becomes quite complicated. To be successful, first, it requires that, not just concrete particulars, but also universals, *have parts*, and that these parts are *also* universals, although not (as for the properties of universals, if such exist) universals of a higher order. A part of a universal (at least this is what Armstrong argues in his 1978b) is not to be equated with a part of the aggregate of particulars which instantiate it. Equally, it is not a sub-class of the class of particulars which instantiate it. Most likely, Armstrong suggests, determinate universals which resemble each other by being partially identical, are *structural* universals, and they are also and simultaneously a certain state-of-affairs type (1978b, 122). Here's Armstrong again (1989a, 106):

> Consider the property of being just five kilograms in mass. For something to have that property the thing must consist of two parts, parts with no overlap between them, such that one part is just four kilos in mass, the other just one. It is a simple form of structural property, simple because no special relations are needed between the two parts: The parts can be scattered parts. We can use the language of states of affairs. The state of affairs of something's being a five-kilo object is the conjunction of two states of affairs: something's being four kilos plus something else's (nonoverlapping something else) being a one kilo state of affairs.

On this view, as we have seen, all universals (at least those able to resemble other universals) must be understood as complexes made up from parts (which are *also* universals). This is relatively easily made sense of if we think of properties like determinate masses, lengths, and durations (which also happen to be Armstrong's favorite examples of properties when defending this view). For these are typically extensive properties. However, not all properties are like that. Some properties are (or appear to be) *intensive* rather than extensive. Colors are one example. These are properties which do not obviously have the complex structure required for the proposal to go through. In his 1978b, Armstrong responds to this worry by pointing out that the reason colors do not appear to be complex is because we judge their nature by their phenomenology. But we shouldn't. The true nature of the colors is revealed to us by science, and science tells us that colors, just like masses, lengths, and durations, are internally structured. Similar arguments are given in Armstrong 1989a, 107 ("the color properties have a concealed complexity") and in Armstrong 1997, 57–61 (where the phenomenology of color is explained by saying that the properties revealed in perception are "second-class"). But this response arguably misses the point, which is that a view that forbids resemblance between simple properties burdens the theory with a whole host of 'extra' theoretical assumptions. Not only in the case of color (which, Armstrong admits explicitly, forces him to regard Physicalism as a 'premise' on which a large portion of his overall metaphysics rests (1997, 58)), but in general. For Armstrong must not only consider color a (disguised) extensive property. He must consider *all* properties – at least all properties which resemble other properties – that way. But there are reasons not to accept this. Electrons, for instance, are often thought of as point-sized. Yet they are also taken to have determinate mass. This means that the following certainly seems possible (not just metaphysically, but also physically): two point-sized particles that have different mass, and that resemble each other 'mass-vise'. But Armstrong must – and does – reject this possibility (Armstrong 1997, 64):

> I take courage and declare that this metaphysics has no place for such quantities. They are as objectionable as determinable/determinate relations holding between simple properties.

In declaring this, however, Armstrong arguably contradicts the basic tenets of his own scientific realism. Or, to put the objection more mildly, in declaring this, Armstrong burdens his account with yet another not further explicable theoretical assumption. An assumption, moreover, that doesn't seem to have any justification other than that it is needed in order for Armstrong's account of resemblance in terms of identity to have general applicability. To require of all properties – that resemble – that they are complex, finally, might makes one wonder about the possible existence of simple properties. Either such properties exist, or they don't. If they do, then those are properties which *cannot* resemble other properties. Which seems a bit weird. And if they don't exist, then *all* properties are complex properties, which means that Armstrong's account automatically – and necessarily – commits him to the existence of, not just possible – but actual – gunky structures. A substantial commitment, indeed.

The price of being able to account for all kinds of resemblance in terms of identity is in other words high. It does not cost you an additional set of primitive axioms of resemblance, mind you. But it's unclear if this 'saving' is enough to tip the balance in the universal realist's favor. For, if resemblance is accepted as primitive, your theory of resemblance at least would seem to become considerably more straightforward: concrete particulars resemble each other because the tropes they instantiate do, and those tropes resemble each other because of their nature. Which, then, is the theoretically simpler view?

***

In this paper, Armstrong's main reasons for rejecting the trope view have been critically scrutinized. And all of them have been found wanting. Does this mean that Armstrong ought to have been a trope theorist? If the 'best' version of the trope view is ontologically more parsimonious than is the 'best' version of the universals view, then yes. That it is has also been argued in this text. Therefore, Armstrong ought to have been a trope theorist.

Acknowledgments: I would like to thank Johan Brännmark for his insightful comments and – not least – for his (as always) unwavering support. I would also like to thank Francesco Calemi for putting this volume together. Most of all I would have liked to thank David Armstrong. A constant source of inspiration. Always challenging. Never dull. Thank you!

Bibliography

Armstrong, D. M. 1978a. *Universals and Scientific Realism, Vol. 1, Nominalism and Realism*. Cambridge: Cambridge University Press.
Armstrong, D. M. 1978b. *Universals and Scientific Realism, Vol. 2, A Theory of Universal*. Cambridge: Cambridge University Press.
Armstrong, D. M. 1980. "Against 'Ostrich Nominalism': a Reply to Michael Devitt". *Pacific Philosophical Quarterly*, 61, 440–449.
Armstrong, D. M. 1983. *What is a Law of Nature?* Cambridge: Cambridge University Press.
Armstrong, D. M. 1989a. *Universals: An Opinionated Introduction*. Boulder, Colorado: Westview Press.
Armstrong, D. M. 1989b. *A Combinatorial Theory of Possibility*. Cambridge: Cambridge University Press.
Armstrong, D. M. 1991. "Classes are States of Affairs". *Mind*, 100(398), 189–200.
Armstrong, D. M. 1997. *A World of States of Affairs*. Cambridge: Cambridge University Press.
Armstrong, D. M. 2004. *Truth and Truthmakers*. Cambridge: Cambridge University Press.
Armstrong, D. M. 2005. "Four Disputes about Properties". *Synthese*, 144(3), 309–320.
Cameron, R. 2006. "Tropes, Necessary Connections, and Non-Transferability". *Dialectica*, 60(2), 99–113.
Campbell, K. 1990. *Abstract Particulars*. Oxford: Basil Blackwell Ltd.
Carroll, J. W. 2012. "Laws of Nature". In Zalta, E. N. (ed.), T*he Stanford Encyclopedia of Philosophy* (Spring 2012 Edition), URL = <http://plato.stanford.edu/archives/spr2012/entries/laws-of-nature>.
Devitt, M. 1980. "'Ostrich Nominalism' or 'Mirage Realism'". *Pacific Philosophical Quarterly*, 61, 433–439.

Dretske, F. 1977. "Laws of Nature". *Philosophy of Science*, 44, 248–268.
Garcia, R. K. 2015. "Is Trope Theory a Divided House?" In Galluzzo, G. and Loux, M. J. (eds.), *Universals in Contemporary Philosophy*. Cambridge: Cambridge University Press, 133–155.
Heil, J. 2003. *From and Ontological Point of View*. Oxford: Clarendon Press.
Hochberg, H. 2001. *The Positivist and the Ontologist: Bergmann, Carnap and Logical Realism*. Amsterdam: Rodopi.
Hochberg, H. 2004. "Relations, Properties, and Predicates". In Hochberg, H. and Mulligan, K. (eds.), *Relations and Predicates*. Heusenstamm: Ontos Verlag, 17–53.
Labossiere, M. C. 1993. "Swapped Tropes". *Pacific Philosophical Quarterly*, 74, 258–264.
Lewis, D. K. 1983. "New Work for a Theory of Universals". *Australasian Journal of Philosophy*, 61(4), 343–377.
Loux, M. J. 2015. "An Exercise in Constituent Ontology". In Galluzzo, G. and Loux, M. J. (eds.), *Universals in Contemporary Philosophy*. Cambridge: Cambridge University Press, 9–45.
Martin, C. B. 1980. "Substance Substantiated". *Australasian Journal of Philosophy*, 58(1): 3–10.
Maurin, A.-S. 2005. "Same but Different". *Metaphysica*, 6(1), 129–145.
Maurin, A.-S. 2008. "The One over Many". In De Mey, T., and Keinänen, M. (eds.), *Problems from Armstrong. Acta Philosophica Fennica*, vol. 84, 37–50.
Maurin, A.-S. 2010, "Trope Theory and the Bradley Regress". *Synthese*, 175(3), 311–326.
Maurin, A.-S. 2013. "Infinite Regress Arguments". In Svennerlind, J., Almäng, J., and Ingthorsson, R. (eds.), *Johanssonian Investigations*. Heusenstamm: Ontos Verlag, 421–438.
Maurin, A.-S. 2014. "Tropes". In Zalta, E. N. (ed.), *The Stanford Encyclopedia of Philosophy* (Fall 2014 Edition), URL = <http://plato.stanford.edu/archives/fall2014/entries/tropes>.
Maurin, A.-S. 2015, "States of Affairs and the Relation Regress", in G. Galluzzo and M. Loux, eds., *The Problem of Universals in Contemporary Philosophy*, Cambridge: Cambridge University Press, 195–214.
Molnar, G. 2003. *Powers: A Study in Metaphysics*. S. Mumford (ed.). Oxford: Oxford University Press.
Mulligan, K., Simons, P., and Smith, B. 1984. "Truth-Makers". *Philosophy and Phenomenological Research*, 44, 287–321.
Simons, P. 1994. "Particulars in Particular Clothing: Three Trope Theories of Substance". *Philosophy and Phenomenological Research*, 54(3), 553–575.
Stout, G. F. 1921. "The Nature of Universals and Propositions". *Proceedings of the British Academy*, 10, 157–172.
Stout, G. F. 1923. "Are the Characteristics of Particular Things Universal or Particular?" *Proceedings of the Aristotelian Society*, supp. vol., 3, 114–122.
Tooley, M. 1977. "The Nature of Laws". *Canadian Journal of Philosophy*, 7, 667–698.
Van Fraassen, B. 1989. *Laws and Symmetry*. Oxford: Clarendon Press.
Williams, D. C. 1953. "On the Elements of Being". *The Review of Metaphysics*, 7(1), 3–18.

William F. Vallicella
Facts: An Essay in Aporetics

1 Introduction

This essay explores a topic central to the work of David M. Armstrong, that of states of affairs or *facts* as I shall call them. In earlier work, much influenced by Armstrong and Gustav Bergmann, I took a realist line, defending concrete facts as the truth-makers of (some) contingently true sentences, and then putting them to work for ambitious metaphysical purposes. (Vallicella 2000, 2002) Since then I have become increasingly aware of how problematic facts are despite their seeming indispensability. I am now tempted by the aporetic conclusion that we cannot live with them and we cannot live without them. Facts are contested entities. Some, such as Panayot Butchvarov, deny their existence altogether. (Butchvarov 1979, 244–247) Others admit them but differ dramatically as to their nature. I see the main division among the friends of facts as that between concretists who locate them in the space-time world and abstractists who don't. I will begin with the concretist conception of facts as contingent truth-makers. This is the conception we find in Armstrong. The next task will be to confront Butchvarov's formidable anti-fact arguments. I then examine some of the problems with the concretist view, and how under dialectical pressure Armstrong came to modify his version of it near the end of his career. This is followed by a look at the abstractist view of facts that we find in Reinhardt Grossmann. Grossmann's could be called a hybrid abstractist view in that he considers first-order facts to have concrete subject constituents. (Grossmann 1992, 73–84) I will not discuss the purely abstractist view of states of affairs one finds in Roderick Chisholm according to which they are "abstract entities which exist necessarily and which are such that some but not all of them occur, take place or obtain." (Chisholm 1976, 114; cf. Plantinga 1974, 44–45) On the purely abstractist view facts are insufficiently different from propositions to warrant discussion here. Of the four main views of facts just distinguished, the eliminativist, the concretist, the hybrid abstractist, and the pure abstractist, I will therefore consider only the first three. I will conclude in good old Platonic fashion, aporetically, as intellectual honesty seems to demand. The road to the impasse, however, should prove instructive.

William F. Vallicella: 5172 South Marble Drive, Gold Canyon, AZ 85219-3387, email: billvallicella@cs.com

2 Facts as contingently existing concreta

As for terminology, I use 'fact' and 'state of affairs' interchangeably, but favor 'fact' on account of its brevity. If facts are truth-makers, however, then we cannot mean by 'fact' what Frege means by *Tatsache*, namely, a true proposition, where a proposition or thought (*Gedanke*) is the sense (*Sinn*) of a context-free declarative sentence. (Frege 1976, 50) Propositions are either true or false, but no fact is either true or false. A proposition is a truth-*bearer*, but a fact is a truth-*maker*. Propositions are bivalent, but there is no corresponding bivalence with respect to facts on the concretist conception. It is not as if some facts obtain and other do not: a fact cannot exist without obtaining. Nor is it the case that some facts are possible and some are actual. A merely possible fact is not a fact. (In formal mode: 'merely possible' in 'merely possible fact' is an *alienans* adjective.) Moreover, on the concretist approach there is no such Meinongian distinction as we find in Reinhardt Grossmann between existent facts and nonexistent states of affairs. (Grossmann 1992, 73) All of this is in keeping with the consideration that facts as truth-makers are ontological grounds 'in the world,' not epistemic or referential intermediaries either 'in the mind' or in Frege's region of senses 'between' mind and world. Fregean propositions reside in a third realm 'between' minds and mental contents (the second realm) and the first realm of primary reference. The concrete facts presently under examination reside in the first realm. They are the referents of sentences, not the senses of sentences. They are concrete (spatiotemporal), not abstract (non-spatiotemporal). They are in the world and play a causal role there. Indeed, for Armstrong, they *are* the world, so much so that the world is one big fact or state of affairs. (Armstrong 1993, 429–440)

An example of a fact is a particular's instantiating a property, or two or more particulars' instantiating a relation. My being happy and my sitting on a rock are examples of facts. On all theories, facts or states of affairs have constituents. A fact is a sort of whole, and its constituents are its (proper) parts, though not in the precise sense of mereology. Thus the fact of *a's being F* has *a* and F-ness as its primary constituents, and the fact of *a's bearing R to b* has *a*, R, and *b* as its primary constituents. I say 'primary' to allow for secondary constituents such as a Bergmannian nexus of exemplification (Bergmann 1967, 9), although some theorists, Armstrong being one of them, deny the need for such a nexus. (Armstrong 1978, 108–111) Note that the fact of *Al's being fat* has Al himself, all 250 lbs of him, as one of its constituents, not a Fregean sense or other abstract surrogate that represents him. Otherwise this fact would not be a truth-maker but would need one.

A fact is a complex, but not every complex is a fact. The set {Al, fatness} is a two-membered complex, but it is distinct from the fact of *Al's being fat*. The existence of the set is entailed by the existence of its members, but the existence of the set and its members does not entail the existence of the fact. The same holds for the extension of the set, if you care to distinguish a set and its extension. The extension of the set does not entail the fact. The same goes for the mereological sum Al + fatness. It can exist without the fact existing. More simply, the existence of a fact's constituents does not entail the existence of the fact. Jack, Jill, and the relation *loves* could each exist without the fact of *Jack's loving Jill* existing. We can express this by saying that the constituents of a fact are externally related to one another: the nature of the constituents does not dictate or necessitate their relatedness. Thus a fact, while composed of its constituents, is more than its constituents: it is their peculiar fact-making togetherness, a togetherness whose peculiarity is that it ties the constituents into a truth-maker.

Every fact has one or more properties as constituents though there is controversy over whether properties are universals or tropes. Bergmann, Armstrong, and Reinhardt Grossmann maintain that properties are universals and that all universals are *immanent* in the sense that they cannot exist unexemplified. Thus for these philosophers they all exist *in rebus* in one sense of this phrase. But one could hold that universals are transcendent in the sense that they can exist unexemplified. It is important to realize that if universals are immanent it does not follow that they are constituents of the things that have them. Immanence and constituency are distinct concepts. For Armstrong, universals are both immanent – cannot exist uninstantiated – and are constituents of the things (thick particulars) that have them. For Grossmann, however, universals, while immanent, are not constituents of the things that have them. *In rebus* (in things), said of universals, is therefore ambiguous: it could mean that universals exist only if instantiated, or it could mean that universals exist only as constituents of things, or both. In Armstrong it means both. It is easy to become confused here since both 'immanent' and 'transcendent' are ambiguous. In my usage, a universal is immanent if and only if it cannot exist uninstantiated, if and only if it is metaphysically necessary that it have at least one instance. But 'immanent' could be used to mean that instantiated universals are 'in' things as their constituents. As for 'transcendent,' in my usage a transcendent universal is one that is metaphysically capable of existing uninstantiated. But it could be used to mean that no universals are 'in' things as their constituents. It follows on my usage of terms that if universals are transcendent, that does not rule out their being constituents of the things that have them. And if universals are immanent, that does not rule in or entail their being constituents of the things that have them. Thus the transcendent/immanent distinction-pair cuts perpendicular to the constituent/nonconstituent distinction-pair.

For present purposes we will take properties and relations to be universals and universals to be immanent. Now consider an ordinary or 'Moorean' particular such as an apple or a round red spot or an electron. Moorean particulars are the items that we all agree pre-analytically are particulars whether or not they are everyday meso-particulars. Are such particulars facts? For Armstrong they are; for Grossmann they aren't. (Cumpa and Tegtmeier 2009, 43–45) For Armstrong, as for Bergmann, an ordinary particular such as a round red spot – an Iowa example! – is a fact. If N-ness is the conjunction of all of the spot's intrinsic (non-relational) properties, then the spot is the fact of *a's being N*. Reality for Armstrong, as for his teacher John Anderson, is sentence-like rather than list-like. (Armstrong 2010, 34) If ordinary particulars are facts, then the particular 'in' such a fact must be a 'thin' particular. What makes a thin particular thin is not that it instantiates no properties, but that it lacks a nature that necessitates that it instantiate certain properties but not others. Thin particulars are 'promiscuous' in the sense that they can connect or combine with any (first-order) property. A thin particular must have some properties or other, but it is contingent which properties it has. All of its properties are accidental. The necessity of having properties goes with the contingency of the properties had. And the same holds for (first-order) universals: they must be instantiated by some particulars or other, but it is contingent which particulars instantiate them. The necessity of being instantiated by particulars goes with the contingency of which particulars instantiate them.

Is there are a good reason to identify Moorean particulars with facts? Suppose that properties are universals and that universals are immanent. That is consistent with the universals being abstract (non-spatiotemporal) objects as on Grossmann's scheme. But if one is a naturalist, if one holds that reality is exhausted by space-time and its contents, then there is no realm of Platonica and the universals that things have must be ontological ingredients or constituents of them. A bundle-of-universals theory would satisfy this requirement. But if, as Armstrong holds, bundle theories are to be rejected, then Moorean particulars are facts.

3 The truth-maker argument for facts

The central and best among several arguments for facts is the Truth-Maker Argument. Take some such contingently true affirmative singular sentence as 'Al is fat.' Surely with respect to such sentences there is more to truth than the sentences that are true. There must be something external to a true sentence that grounds its being true, and this external something is not plausibly taken to be another sentence or the say-so of some person. 'Al is fat' is not just true; it is true *because* there is something in extralinguistic and extramental reality that 'makes' it true, something 'in virtue of which' it is true. There is this short man, Al, and the guy weighs 250 lbs. There is nothing linguistic or mental about the man or his weight. Here is the sound core, at once both ancient and perennial, of correspondence theories of truth. Our sample sentence is not just true; it is true because of the way the world outside the mind and outside the sentence is configured. The 'because' is not a causal 'because.' The question is not the empirical-causal one as to why Al is fat. He is fat because he eats too much. The question concerns the ontological ground of the truth of the sentential representation, 'Al is fat.' Since it is obvious that the sentence cannot just be true – given that it is not true in virtue of its logical form or *ex vi terminorum* – we must posit something external to the sentence that 'makes' it true. I don't see how this can be avoided even though I cheerfully admit that 'makes true' is not perfectly clear. That (some) truths refer us to the world as to that which makes them true is so obvious and commonsensical and indeed 'Australian' that one ought to hesitate to reject the idea because of the undeniable puzzles that it engenders. Motion is puzzling too but presumably not to be denied on the ground of its being puzzling.

Now what is the nature of this external truth-maker? *Truth-maker* is an office. Who or what is a viable candidate? It can't be Al by himself, if Al is taken to be ontologically unstructured, an Armstrongian 'blob,' as opposed to a 'layer cake' and it can't be fatness by itself.[1] (Armstrong 1989a, 38, 58) If Al by himself were the truth-maker of 'Al is fat' then Al by himself would make true 'Al is not fat' and every sentence about Al whether true or false. If fatness by itself were the truth-maker, then fatness exemplified by some other person would be the truth-maker of 'Al is fat.' Nor can the truth-maker be the pair of the two. For it could be that Al exists and fatness exists, by being exemplified by Sal, say, but Al does not instantiate fatness. What is needed, apparently, is a proposition-like entity, the fact of *Al's being fat*. We need something in the world to undergird the predicative tie. So it seems we must add the category of fact to our ontology, to our categorial inventory. *Veritas sequitur esse* – the principle that truth follows being, that there are no truths about what lacks being or existence – is not enough. It is not enough that all truths are about existing items *pace* Meinong. It is not enough that 'Al' and 'fat' have worldly referents; the sentence as a whole needs a worldly referent. In many cases, though perhaps not in all, truth-makers cannot be 'things' – where a thing is either an individual or a property – or collections of same, but must be entities of a different categorial sort. Truth-making facts are therefore 'an addition to being,' not 'an ontological free lunch,' to employ a couple of signature Armstron-

---

[1] If Al is a blob, then he lacks ontological structure; but that is not to say that he lacks spatial or temporal parts. It is obvious that he has spatial parts; it is not obvious that he has ontological 'parts.' Thin particulars, properties, and nexus count as ontological 'parts.' Layer cakes have both spatiotemporal and ontological structure.

gian phrases. For the early Armstrong at least, facts do not supervene upon their constituents. This yields the following scheme. There are particulars and there are universals. The Truth-Maker Argument, however, shows or at least supports the contention that there must also be facts: particulars-instantiating-universals.[2] There are other arguments for facts, but they cannot be discussed here. And there are other candidates for the office of truth-maker such as tropes and Husserlian moments (Mulligan et al. 2009) but these other candidates cannot be discussed here either. Deeper than any particular argument for facts, or discussion of the nature of facts, lies the question whether realism about facts even makes sense. To this question we now turn.

4 Butchvarov's objections to realism about facts

The Truth-Maker Argument for facts is impressive, but the very notion of a fact, regardless of the arguments given for their admission, give rises to puzzles and protests. There is the Strawsonian protest that facts are merely hypostatized sentences, shadows genuine sentences cast upon the world. Panayot Butchvarov sympathetically quotes P. F. Strawson's seminal 1950 discussion: "If you prise the sentences off the world, you prise the facts off it too..." (Butchvarov 2010, 73–74; Strawson 1950) Strawson again: "The only plausible candidate for what (in the world) makes a sentence true is the fact it states; but the fact it states is not something in the world." Why aren't facts in the world? This section considers two formidable but inconclusive arguments.

4.1 An argument from imperceivability

'The table is against the wall.' This is a true contingent sentence. I know that it is true by seeing (or otherwise sense perceiving) that the table is against the wall. This seeing is arguably the seeing of a fact, where a fact is not a true proposition

---

[2] Are facts or states of affairs then a third category of entity in addition to particulars and universals? Armstrong fights shy of this admission: "I do not think that the recognition of states of affairs involves introducing a new entity [...] it seems misleading to say that there are particulars, universals, *and* states of affairs." (Armstrong 1978, 80) Here we begin to glimpse the internal instability of Armstrong's notion of a state of affairs. On the one hand, it is something in addition to its constituents: it does not reduce to them or supervene upon them. On the other hand, it is not a third category of entity. We shall see that this instability proves disastrous for Armstrong's ontology.

but the truth-maker of a true proposition. This seeing is not the seeing of a table (by itself), nor of a wall (by itself), nor of the pair of these two physical objects, nor of a relation (by itself). It is the seeing of a table's standing in the relation of being against a wall. It is the seeing of a truth-making fact. So it seems we have here an argument for adding facts to the categorial inventory. The relation, however, is not visible, as are the table and the wall. So how can the fact be visible, as it apparently must be if I am to be able to see (literally, with my eyes) that the table is against the wall? If the relation ingredient in a relational fact is invisible, how can the fact be visible? This is a problem Butchvarov poses for us. Let 'Rab' symbolize a contingent relational truth about observables such as 'The table is against the wall.' We can then set up the problem as an aporetic pentad:

1. If one knows that Rab, then one knows this by seeing that Rab (or by otherwise sense-perceiving it).
2. To see that Rab is to see a fact.
3. To see a fact is to see each of its constituents.
4. The relation R is a constituent of the fact that Rab.
5. The relation R is not visible (or otherwise sense-perceivable).

The pentad is inconsistent: the conjunction of any four limbs entails the negation of the remaining one. To solve the problem, then, we must reject one of the propositions. But which one?

(1) is well-nigh undeniable: I sometimes know that the cat is on the mat, and I know that the cat is on the mat by seeing that he is. How else could I know that the cat is on the mat? I could know it on the basis of the testimony of a reliable witness, but then how would the witness know it? Sooner or later there must be an appeal to someone's direct seeing. (5) is also undeniable: I see the cat; I see the mat; but I don't see the relation picked out by 'x is on y.' And it doesn't matter whether whether you assay relations as relation-instances or as universals. Either way, no relation appears to the senses.

Butchvarov in effect denies (2), thereby converting our pentad into an argument against facts, or rather an argument against facts about observable things. (Butchvarov 2010, 84–85; Butchvarov 1979, 244) The perceivability of relational facts involving observables stands and falls with the perceivability of relations, but relations, Butchvarov insists, are not perceivable. But if there are no facts about observable things, then it is reasonable to hold that there are no facts at all. So one solution to our problem is the 'No Fact Theory.'

One problem I have with Butchvarov's denial of facts is that (1) seems to entail (2). Now Butchvarov in effect grants (1). So why doesn't he grant (2)? In other words, if I can see (with my eyes) that the cat is on the mat, why isn't that excellent evidence that I am seeing a fact and not just a cat and a mat? If you grant me that I sometimes see that such-and-such, must you not also grant me that I sometimes see facts? I am seeing something, and it is not a sentence or a shadow cast by a sentence. Nor am I seeing a cat, a mat, and a relation. I am seeing a cat's *being* in a familiar relation to a mat. And if there are no facts, then how do we explain the truth of contingently true sentences such as 'The cat is on the mat'? As explained above, there is more to the truth of this sentence than the sentence that is true. The sentence is not just true; it is true because of something external to it, something which, though not a proposition, is proposition-like.

Another theory arises by denying (3). Butchvarov would not find this denial plausible. If I see the cat and the mat, why can't I see the relation – assuming that I am seeing a fact and that a fact is composed of its constituents, one of them being a relation? As Butchvarov asks, rhetorically, "If you supposed that the relational fact is visible, but the relation is not, is the relation hidden? Or too small to see?" (Butchvarov 2010, 85)

A third theory comes of denying (4). One might deny that R is a constituent of the fact of *a's standing in R to b*. But surely this theory is a nonstarter. If there are relational facts, then relations must be constituents of some facts. Our problem seems to be insoluble. Each limb makes a very strong claim on our acceptance. But they cannot all be true. Butchvarov has not shown compellingly that there are no facts, but he has cast serious doubt upon them.

4.2 An argument from impossibility of reference

Perhaps the weakest argument Armstrong gives for facts is that we can refer to them and what we can refer to exists. (Armstrong 1989a, 89) But *can* we refer to them? Butchvarov thinks not. In his essay, "Facts," (Butchvarov 2010) Panayot Butchvarov generously cites me as a defender of realism and a proponent of facts. He credits me with doing something William P. Alston does not do in his theory of facts, namely, specifying their mode of reality:

> However, William Vallicella, also a defender of realism, does. He argues that true propositions require "truth-making facts." And he astutely points out that facts could be truth-making only if they are "proposition-like," "structured in a proposition-like way" – only if a fact has a structure that can mirror the structure of a proposition. (Vallicella 2002, 13, 166–7, 192–3) Vallicella's view is firmly in the spirit of Wittgenstein's account in the *Tractatus* of the notions of fact and correspondence to fact, but his formulation of it may invite deflationist attacks like Strawson's.

Butchvarov, however, is firmly against adding the category of facts to our ontological inventory. Butchvarov tells us (Butchvarov 2010, 86) that

> The metaphysical notion of fact is grounded in our use of declarative sentences, and the supposition that there are facts in the world depends at least in part on the assumption that sentences must correspond to something in the world, that somehow they must be names. But this assumption seems absurd. Sentences are not even nouns, much less names. They cannot serve as grammatical subjects or objects of verbs, which is the mark of nouns. [...] Notoriously, "p is true," if taken literally, is gibberish. "Snow is white is true" is just ill-formed. "'Snow is white' is true" is not, but its subject-term is not a sentence – it is the name of a sentence.

Here is what I take to be Butchvarov's argument in the above passage and surrounding text:

1. If there are facts, then some declarative sentences are names.
2. Every name can serve as the grammatical subject of a verb.
3. No declarative sentence can serve as the grammatical subject of a verb.
   Therefore
4. No declarative sentence is a name. (2, 3)
   Therefore
5. There are no facts. (1, 4)

The friend of facts ought to concede (1). If there are truth-making facts, then some declarative sentences refer to them, or have them as worldly correspondents. The realist holds that if a contingent sentence such as 'Al is fat' is true, then that is not just a matter of language, but a matter of how the extralinguistic world is arranged. The sentence is true because of *Al's being fat*. As for (2), it is undeniable. So if the argument is to be neutralized we must give reasons for not accepting (3). The trouble with (3) is that it is deeply paradoxical. It implies that the sentence 'Snow is white' is not a sentence and therefore cannot be true!

4.3 The Paradox of the Horse and the Paradox of Snow

Butchvarov is committed to holding that a sentence like 'Snow is white' is not a sentence but the name of a sentence. The paradox is similar to the paradox of the horse in Frege. (Frege 1960, 46) Frege notoriously holds that the concept *horse* is not a concept. Butchvarov is maintaining that the sentence 'Snow is white' is not a sentence. This paradox engenders others. If 'Snow is white' is not a sentence, but a name, then it is neither true nor false. But it is obviously true, not false. And if 'Snow is white' is not a sentence, then how does it differ from 'snow' which is obviously not a sentence?

What is Frege's reasoning? He operates with a mutually exclusive distinction between names and predicates (concept words). No name is a predicate and no predicate a name. Corresponding to this linguistic distinction there is the mutually exclusive ontological distinction between objects and concepts. No object is a concept, and no concept an object. Objects are nameable while concepts are not. So if you try to name a concept you cannot succeed: willy-nilly you transform it into an object. Since 'the concept *horse*' is a name, its referent is an object. Hence the concept *horse* is not a concept but an object. It follows that one cannot say that the concept *horse* has instances. For that would be to say, nonsensically, that an object has instances. Nor can one say, meaningfully, that the concept *horse* includes the concept *mammal*. But it seems pretty clear that one can say that both meaningfully and with truth.

Similarly with Butchvarov. He operates with a mutually exclusive distinction between names and sentences. No name is a sentence, and no sentence a name. To refer to a sentence, I must use a name for it. To form the name of a sentence, I enclose it in quotation marks. Thus the sentence 'Snow is white' is not a sentence, but a name for a sentence. But then, since names are neither true nor false, 'Snow is white' is neither true nor false. Butchvarov thinks we face a dilemma. 'Snow is white' is not true, but 'Snow is white is true' is gibberish. Butchvarov finds it "absurd" that a sentence should name a fact. (Butchvarov 2010, 86) His reason is that a sentence is not a name. But it is equally or even more absurd to say that the sentence 'Snow is white' is not a sentence, but a name. For the sentence in question is true, and no name is either true or false.

Against both Frege and Butchvarov I would say that it is not at all clear that we must make mutually exclusive distinctions between objects and concepts and between names and sentences. The essence of a concept is its predicability, but there is no compelling reason why an item cannot be both predicable and a subject of predication. Thus *horse* can be both predicated and made the subject of predication as when I say 'The concept *horse* includes the concept *mammal*.' Similarly with Butchvarov. It is not clear why we cannot say that some sentences can function as names without ceasing to be sentences. 'Snow is white' is both a sentence and a name. It is a sentence that names a fact. My tentative conclusion is that while realism about facts is dubious, so is Butchvarov's rejection of realism. Premise (3) in the above argument is rendered suspect by its paradoxicality.

5 Problems with the concretist conception of facts

Butchvarov's objections apply to facts on both the concretist and abstractist conceptions. The concretist conception gives rise to puzzles of its own. This section will address only some of them. The focus will be on Armstrong's concretism.

5.1 The collision of the compositional and necessitarian models

Facts have constituents and are nothing without them. No fact is simple, all are complex. Facts are wholes of parts, albeit in stretched senses of 'whole' and 'part.' (Armstrong 2010a, 32) In the first-order cases, whether monadic or polyadic, facts unify members of two mutually irreducible categories, particulars and universals. The categories are mutually irreducible because particulars cannot be assayed as bundles of compresent universals, and because universals cannot be reduced to particulars or to classes of particulars such as classes of resembling tropes. This suggests a building-block or *compositional model*: facts are built up out of ontologically more basic materials belonging to irreducibly different categories. Atomic facts are composed of particulars and universals. Of course, this building-up or composing is not a temporal process. Particulars and universals are not temporally prior to facts, but ontologically prior. Here is a crude analogy. Suppose that stone walls consisting of appropriately stacked stones and nothing else always existed: there was never a time when the stones composing a given wall were not arranged wall-wise. There was never a time when the stones were just laying about. We would nevertheless consider the stones to be the basic materials out of which the walls are constructed. Not that a wall is just stones: it is stones arranged wall-wise. The point is that the stones, even if always parts of walls, would be basic relative to walls. Furthermore, if stone s is a proper part of wall W, we would not say that s's nature dictates that it belong to W as opposed to W*. If there were 1000 stones total and ten walls each composed of 100 stones, we would not say that the ten actual wall-wise arrangements were the only combinatorially possible ones.

Armstrong's initial fact model is compositional and combinatorial: there is a stock of particulars and a stock of universals and these give rise to a set of possible combinations only some of which constitute facts. Suppose there are two particulars, a, b, and two monadic universals, F-ness, G-ness. Then there are four combinatorially possible combinations: a + F-ness, a + G-ness, b + F-ness, b + G-ness. But it may be that there are only two facts: *a's being F* and *b's being G*. The other two combinations are not facts in the mode of mere possibility: they are not facts at all. All facts exist (obtain, are actual). But not all existents are facts: the subfactual constituents of facts exist but are not facts. This yields three categories of existent: particulars, universals, and facts, with members of the third category composed of members of the other two.

Facts on the compositional model are an addition to being: they do not supervene upon their constituents. They are not a 'free lunch' ontologically speaking. This is why not all possible combinations of fact-friendly items are facts. If a exists and F-ness exists, it does not follow that *a's being F* exists. So not every possible combination of a particular and a universal constitutes a fact. Not every such combination is one in which the particular instantiates the universal. Instantiation is contingent and non-supervenient. And yet Armstrong insists that facts "are primary, particulars and universals secondary." He takes this to mean that facts "are the least thing that can have *independent* existence." He goes on to say that particulars and universals are "false abstractions," "incapable of independent existence." (Armstrong 2010a, 27) Armstrong also speaks of "impossible abstractions" (Armstrong 2009, 42) and "illegitimate abstractions."

But here we face a deep unclarity. Does Armstrong mean to say that whatever exists exists independently, that particulars and universals do not exist independently, and that therefore, particulars and universals do not exist? Is that what is meant by talk of their being false or impossible abstractions? This seems highly unlikely. If particulars and universals do not exist, then facts, which exist, cannot be composed of them. Surely that is blindingly obvious. Facts may be unmereological compositions, but they are compositions of sub-factual elements. And yet puzzles lurk beneath the surface. In his book, *David Armstrong*, Stephen Mumford correctly notes that for Armstrong atomic facts are the "fundamental entities of the world," "the smallest possible units of existence." (97) They are "independent existences." (96) But of course facts have constituents: particulars, properties, and relations. Mumford tells us that "[...] while these are real enough, they are not themselves existents." (97) The particulars and universals within atomic facts, whether monadic or polyadic, are abstractions from what exists, namely, the facts. The analysis of an atomic fact into its sub-factual components is "by abstraction, not by any real process." (96) This is puzzling: how can the sub-factual constituents of a fact be real without existing? The puzzle as an aporetic triad:

1. To exist is to be real.
2. Facts alone exist: particulars, properties, and relations do not exist, being abstractions from what exists.
3. Particulars, properties, and relations are real.

The triad is plainly inconsistent. Much of what Armstrong writes suggests that the solution is to reject (2). Facts exist and their constituents exist. It is just that the constituents cannot exist apart from facts. That is equivalent to saying that particulars cannot exist without instantiating universals, and (first-order) universals cannot exist without being instantiated by particulars. But note that the impossibility is a general one and does not extend to the tie between any particular particular and any particular universal. If it did, Armstrong's combinatorial theory of possibility would be impossible. Let me explain.

In the fact of *a's being F*, there is no necessity that a instantiate F-ness or that F-ness be instantiated by a. The only necessity is that a have some property or other and that F-ness be instantiated by some particular or other. It is therefore possible that the constituents of a fact exist without the fact existing. What makes this possible is the circumstance that nothing about a dictates which properties it has and nothing about F-ness dictates which particulars instantiate it. So a and F-ness are independent of a's being F in that the identity and existence of the constituents does not depend on the identity and existence of the fact. F-ness is F-ness regardless of which particulars instantiate it, and a is a regardless of which universals it instantiates. F-ness, after all, is a universal, a one-in-many. As such, its existence and identity are not exhausted by its constituency in any particular fact such as *a's being F*. It is the same universal in *b's being F*. And the same goes for the particular a. Although it is not a one-in-many, being unrepeatably and irreducibly particular, its 'promiscuity' ensures that its existence and identity are not exhausted by its being the particularity of any particular fact. Although a is not G, it might have been. The identity of the fact, however, does depend on the identity of the constituents: each fact has essentially the constituents it has. This fact-essentialism is an analog of mereological essentialism. Facts depend for their identity and existence on their constituents, but constituents do not depend for their identity and existence on the facts into which they enter. There is an asymmetry here. A fact cannot gain or lose constituents or have different constituents in different possible worlds. But a universal, even though it must be instantiated, can be instantiated by different particulars at different times and in different worlds. A particular, though necessarily such as to have properties, can gain or lose properties and have different properties in different worlds.

I have just sketched the position Armstrong holds in most of his writings. But why the puzzling talk about false and impossible and illegitimate abstraction? It is widely accepted even among ontologists who reject facts that there are no unpropertied particulars and no uninstantiated universals. It would make good sense to speak of unpropertied particulars and uninstantiated universals as false or impossible or illegitimate or vicious abstractions. Indeed, Armstrong speaks of them as vicious abstractions in his book on the laws of nature where he distinguishes vicious from non-vicious abstractions. (Armstrong 1983, 84) But a thin particular is not an unpropertied particular. What makes a thin particular thin is not its lacking properties but the manner in which it has the properties it *must* have. Thin particulars are necessarily such as to have properties, though it is contingent *which* properties they have. And an immanent universal is not a universal that lacks instances. Immanent universals are necessarily such as to have instances, though it is contingent *which* instances they have.

It makes no sense, therefore, to speak of thin particulars and immanent universals as false or impossible abstractions on the Compositional or Building Block model. To the extent that Armstrong (and Mumford) speak in this way, they are responding to dialectical pressure from a competing fact model which we can call the Necessitarian or Abstractionist (not abstractist) model. On the Necessitarian model, the general necessity that particulars instantiate universals, and that (first-order) universals be instantiated by particulars becomes a particular necessity within each fact. Thus in *a's being F*, *a* cannot exist except as a constituent of that fact: *a* is necessarily tied to F-ness. And F-ness is necessarily tied to *a* and to every other particular that instantiates it. On the Necessitarian approach, it does make sense to speak of sub-factual constituents as false or impossible or illegitimate abstractions. For on the Necessitarian model, the sub-factual constituents enjoy no independence of the facts in which they are constituents. On the Compositionalist model, by contrast, the sub-factual constituents retain a certain independence in that it is possible that *a*, which is F, but not G, not be F but be G instead, and it is possible that F-ness, which is instantiated by *a* and *b* but not *c* and *d* be instantiated by *c* and *d* but not *a* and *b*.

5.2 Problems with the compositionalist model

What I will now argue is that the Compositionalist model is deeply problematic and carries within it the seeds of its own destruction. It faces the Antinomy of Bare Particulars. Armstrong's solution to the antinomy in terms of his distinction between the thin and the thick particular, however, leads to an antinomy of predication I will call Aristotle's Revenge.

In *Universals: An Opinionated Introduction*, Armstrong discusses a problem John Quilter calls the "Antinomy of Bare Particulars." (Armstrong 1989, 95) Suppose *a* is F. The 'is' is not the 'is' of identity. It expresses the asymmetrical and external tie of instantiation: the particular *a* instantiates the universal F-ness. If so, *a* considered in itself is bare of properties. It is outside and other than all of its properties. But then *a* does not have the property F-ness and *a* is *not* F.

Armstrong attempts to defuse the above contradiction in the time-honored manner, by making a distinction. He distinguishes the thin from the thick particular. The thin particular is the particular "taken apart from its properties (substratum)." (Armstrong 1989, 95). It is linked to its properties by instantiation. The thick particular is the particular taken together with all its nonrelational (intrinsic) properties. Thick particulars "enfold" both thin particulars and the properties they instantiate. On this scheme, thick particulars are facts, plenary facts if you will. Let N ('N' for 'nature') be the conjunction of all of Socrates' nonrelational properties. Then Socrates = the plenary fact of *a's being N*. Socrates' being male, by contrast, is a non-plenary fact. This is my terminology, not Armstrong's.

With this thick versus thin distinction the above antinomy can be solved. We need to avoid the contradiction that *a* is F and *a* is not F. We can say this: the thick particular A is F while the thin particular *a* in A is not F. Thus thick Socrates is wise in virtue of having both thin Socrates and wisdom as constituents with thin Socrates' instantiating wisdom. Thick Socrates is characterized by wisdom but does not instantiate wisdom; thin Socrates instantiates wisdom, but is not characterized by wisdom. Problem solved. Unfortunately, Armstrong's solution give rise to a puzzle of its own.

If Socrates is a fact, and Socrates is seated, then some plenary fact, the fact of *a's being N*, is seated, where 'N' picks out the conjunction of Socrates' non-relational universals. Although it sounds very strange to say of a fact that it is seated, or wise, or sunburned, or in one place rather than another, this strangeness by itself does not amount to an objection. But if Socrates is seated, then this is contingently the case. For while he must be in some posture or other, there is presumably no necessity that he be seated. Now if Socrates is a fact, then the fact in question, *a's being N*, either instantiates the property of being seated or it contains this property as a constituent. But the fact cannot instantiate the property. For it is thin particulars that instantiate properties and by so doing constitute atomic facts. So the fact contains the property. But then the contingency of Socrates' being seated is lost. This because an analog of mereological essentialism holds for facts. Call it fact-essentialism: if F is a fact, and *c* is a constituent of F, then, necessarily, *c* is a constituent of F. A fact cannot gain or lose constituents or have different constituents in different possible worlds on pain of ceasing to be the very fact that it is. A fact's constituents are essential to it. So if Socrates is seated at time *t*, then necessarily Socrates is seated at *t*, contrary to the contingency datum with which we began. The antinomy should come as no surprise given that the relation of a fact to its constituents is a type of whole-part relation.

We could call this antinomy of predication Aristotle's Revenge. Thin particulars are not Aristotelian substances. They do not have natures or essences. That is what their thinness consists in. An ontological scheme in which thin particulars instantiate universals is one in which the nexus of instantiation is and must be external. This implies that all properties are had accidentally and none essentially. But if the particulars we encounter in experience are thin, then we face Quilter's Antinomy of the Bare Particular. Particulars and universals are 'outside' each other and particulars considered in themselves are bereft of intrinsic properties. Thus an apple which is intrinsically (as opposed to relationally) red, if taken to be thin, is not intrinsically red, but only relationally red in virtue of standing in an external exemplification relation to a universal. This difficulty motivates Armstrong's suggestion that ordinary particulars are not mere or pure particulars, but facts or states of affairs composed of a particularizing factor – the thin particular – and a nature factor, the universals. But now Aristotle gets his revenge. Essences were banished, but now it turns out that facts, as non-mereological wholes, have their nature constituents, the universals, essentially! Particulars, thick particulars, have essences after all. It is of the essence of thick Socrates, if seated, to be seated.

5.3 Necessitarianism and the collapse of Armstrong's fact ontology

Near the end of his career, Armstrong radically modifies his conception of facts or states of affairs. The modification is so drastic as to amount to an elimination of his original conception. As we have seen, Armstrong's initial approach is compositional and combinatorial. There are universals and there are particulars and they are mutually irreducible. Nominalism is out and so is what Armstrong dubs 'universalism,' according to which particulars are bundles of universals. (Armstrong 2004, 140) But of course universals are somehow tied to particulars and this tie is real, not merely logical or conceptual. Given this real-world tie, Armstrong finds the acceptance of facts "inevitable" since a (first-order) fact is just a particular instantiating a monadic universal or an n-tuple of particulars instantiating an n-adic universal. (Armstrong 2009, 39) Given naturalism, the view that reality is exhausted by the space-time system, universals must be denizens of space-time. But then they cannot be abstract objects residing in a realm apart from space-time but must be ontological constituents of the things whose universals they are. It is not enough that universals be immanent (existent only if instantiated); they must also literally reside as constituents in spatiotemporal things. Hence instantiation cannot be a chasm-spanning relation as it is for Grossmann, connecting the timeless to the time-bound, but must be internal to things even if it is not a separate constituent. The upshot is that ordinary particulars get construed as concrete facts. Ordinary particular A is identified with the fact of *a's being N*, where 'N' picks out a conjunctive universal the conjuncts of which are A's non-relational universals. But then the postulation of thin particulars becomes inevitable. For it is not A that instantiates universals, but the thin particular a in A. Thin particulars must be thin, i.e. natureless, since they are the instantiators of properties external to them. The thinness of thin particulars entails the contingency of their connection to properties. The instantiation tie, being external, must be contingent. The contingency of the connection entails the promiscuous combinability of particulars and universals, which is essential to Armstrong's combinatorial theory of possibility. (Armstrong 1989, 47) It also entails a certain independence of a thin particular from the facts it is a constituent of, though not from facts in general. No thin particular and no universal can occur outside of a fact, but any such particular can combine with any first-order universal. Bertrand Russell could have been a poached egg, or rather the individuals that constitute Russell could, collectively, have had properties that would have made of those individuals parts of a poached egg. (Armstrong 1989, 51-53)

Armstrong backtracks on almost all of this as his earlier approach gives way to one that can be called necessitarian or 'abstractionist.' But my terminology may mislead. Armstrong's facts are and remain concrete (spatiotemporal). His facts are not 'abstract objects' as are Grossmann's. But while Armstrong's approach is concretist and not abstractist, it ceases to be compositionalist and becomes *abstractionist*. This abstractionism of the later Armstrong involves four major interconnected innovations. First of all, instantiation is assimilated to partial identity. (Armstrong 2004, 139) Second, and in consequence of the first innovation, the link between particulars and non-relational properties (universals) is no longer seen as contingent but as necessary. (Armstrong 2009, 41) Third, Armstrong decides that "The thin particular is an impossible abstraction." (Armstrong 2009, 42) This amounts to a rejection of thin particulars. His earlier view was that, while there are no unpropertied particulars, there are thin particulars that are in no way mental or unreal despite their being abstractions from facts. His talk of abstraction was meant to convey merely that thin particulars and universals "have no existence outside states of affairs." (Armstrong 1989, 43) His earlier view was that while unpropertied or 'bare' particulars are vicious abstractions, thin particulars are non-vicious abstractions. They exist outside the mind but they cannot exist outside of facts. Fourth, Armstrong comes to the view that facts supervene on their constituents. Earlier he held that constituents supervene on facts, but not vice versa. His final view is that the supervenience is symmetrical. (Armstrong 2009, 43) This symmetry of supervenience entails that a is necessarily F if a is F at all.

These innovations, though motivated by problems with the earlier view, are disastrous and as far as I can see they completely undermine Armstrong's fact ontology. Let's start with the assimilation of instantiation to partial identity. This assimilation is motivated by the difficulty of understanding the link in reality between a particular and a universal it instantiates. On the one hand, instantiation is not identity: the 'is' is '*a* is F' is not the 'is' of identity but the 'is' of predication. On the other hand, *a* and F-ness are not wholly distinct existences. If they were, then, Armstrong thinks, the only way to connect them would be via instantiation construed as an external relation. But then Bradley's regress is up and running.[3] So Armstrong, following Donald Baxter, proposes that we think of instantiation as partial identity: particular and universal intersect or overlap. (Baxter 2001, 449–464) It seems clear that if *a* and F-ness are either wholly or partially identical, then no problem could arise as to their connection. Armstrong continues to reject the view that a particular is nothing more than a bundle of universals, and continues to uphold the substance-attribute view, insisting on an "ineliminable factor of particularity in particulars." (Armstrong 2004, 140) "There is something that has the properties..." but on the new view it cannot be a thin particular. Thin particulars are pure particulars that totally exclude universals, just as universals on the old conception totally exclude particulars. (Armstrong 2004, 143) The categories are disjoint and mutually irreducible. But if particulars and universals are partially identical, if the former participate in and overlap the latter, then thin particulars cannot be what have properties. We cannot conclude, however, that thick particulars have properties. For having is instantiation, and thick particulars do not instantiate properties, they contain them as constituents.

Armstrong faces a dilemma. Either universals are partially identical to thin particulars or to thick particulars. They cannot be partially identical to thin particulars because the latter are pure particulars lacking natures. As such, they totally exclude universals, hence cannot overlap them. How could something wholly nonqualitative overlap something that is essentially qualitiative such as the property of being blue? Would the overlap be nonqualitative or qualitative? Universals *can* be partially identical to, and overlap, thick particulars in a manner analogous to the way a part of a whole overlaps the whole in mereology. For a part of a whole, whether proper or improper, does not exclude the whole of which it is a part, and vice versa. But universals are not instantiated by thick particulars any more than a part of a whole is instantiated by the whole. Thick particulars do not instantiate universals; they have them as constituents, as non-mereological parts. So while one can appreciate the dialectic that leads away from instantiation as an external tie to instantiation as partial identity, it appears that Armstrong

---

[3] For a detailed discussion of Bradley's Regress and various responses to it, see Vallicella 2002, 195–239.

has merely traded the Bradley problem for the just-mentioned dilemma. But this is just the beginning of his troubles.

We also note that the rejection of thin particulars, which are not to be confused with unpropertied particulars, brings with it the rejection of thick particulars. That is to say, an ordinary or Moorean particular such as an apple or an electron cannot be conceptualized as a thick particular if there are no thin particulars. The thin and the thick particular define each other. A thick particular is a thin particular together with, instantiating, its non-relational universals, and a thin particular is a thick particular apart from these universals. So without thin particulars there can be no thick particulars. But a thick particular is a concrete fact. A thick particular is not a bundle of compresent universals, but a non-mereological whole in which one of the parts, the thin particular, instantiates the other parts, the universals. So if there are no thin particulars, then Moorean particulars cannot be construed as facts. But then Armstrong's whole ontology collapses. One of his central ideas is that the world, which for him is just the physical universe, is a world of states of affairs or facts. It is not a world of thin particulars whose universals are outside of space-time, not is it a world of bundles, whether of universals or of tropes.

The rejection of thin particulars as impossible abstractions also brings with it the rejection of instantiation, strictly understood. Instantiation as holding between particulars and universals is asymmetric: if a instantiates F-ness, then F-ness does not instantiate a. (Instantiation is not in general asymmetric, but non-symmetric: if one universal instatiates a second, it may or may not be the case that the second instantiates the first.) Partial identity, however, is symmetric: if a overlaps F-ness, then F-ness overlaps a. So, given that instantiation as holding between particulars and universals is asymmetric, it cannot be understood as partial identity or intersection or overlap. This is an argument as powerful as it is simple, and powerful in part because it is so simple and luminous to the intellect.

Armstrong, aware of the difficulty, attempts to accomodate the undeniable asymmetry involved in a particular's instantiating a universal by appealing to the "categorial difference" between particulars and universals and claiming that the asymmetry supervenes upon this difference. (Armstrong 2004, 146) But this move is unavailing. For what does the categorial difference consist in if not the difference between instantiable entities and non-instantiable entities? All and only universals are instantiable; all and only particulars are non-instantiable. Instantiability *constitutes* the categorial difference; it does not merely supervene upon it. But then instantiation cannot be partial identity. I don't see that calling it "non-mereological partial identity" helps. (Armstrong 2004, 141) It is clear that a fact is not a mereological sum of its constituents. It is nonetheless a complex object. So Armstrong concludes that mereological composition is not the only kind of composition. Having convinced himself that instantiation is partial identity, he infers that it too is "non-mereological." (Armstrong 2004, 142) This may be granted, assuming we know what it means, but it does nothing to remove the symmetry of partial identity or to blunt the force of the simple and powerful objection raised in the preceding paragraph. And do we know what it means? I know what it means to say that the fact of *a's being F* is composed of a and F-ness but is more than the mere sum of the two. And I know what it means to say that a instantiates F-ness. I also know what it would mean if someone were to say that instantiation, as a real tie, is non-mereological in that its holding between a and F-ness generates a non-mereological whole. But I have no idea what it could mean to say that partial identity or overlap or intersection are non-merelogical unless I presuppose the existence of facts as non-mereological compositions and confusedly transfer the property of being non-mereological from the fact to the partial identity that is supposed to bring together the constituents. It is some such confusion as this to which Armstrong seems to succumb. Facts are non-mereological compositions because the constituents are contingently connected; but if the connection is partial identity, then the connection is necessary. Talk of partial identity as non-mereological seems just confused.

Armstrong cites the repeatability of universals as a mark of their difference from particulars. (Armstrong 2004, 147) But repeatability in the relevant sense is just multiple instantiability, which brings us back to the asymmetry of instantiation. There is of course a sense in which particulars are repeatable. One and the same particular a is repeated in the facts Fa, Ga, Ha, Rab. But this is different from the sense in which F-ness is repeated in Fa, Fb, Fc. Multiple instantiability is not the same as being the subject of multiple attributes. Armstrong also points to Instantial Invariance as distinguishing universals from particulars. (Armstrong 2004, 147) Particulars are not instantially invariant. There is nothing to stop a from entering into the following facts: $Fa, Rab, Sabc, Tabcd$, etc. Particulars can connect with universals of different 'adicities.' But there is no universal U such that $Ua, Uab, Uabc$. Universals are instantially invariant in that they are either exclusively monadic, dyadic, triadic, etc. Let us grant that there is this difference between universals and particulars. But it cannot be the difference that makes the difference between universals and particulars. Obviously the difference is grounded much deeper in the difference between instantiable and non-instantiable entities and thus in the asymmetry of instantion. It is therefore not a way of distinguishing universals and particulars that is independent of the asymmetry of instantiation. I think we ought to conclude that Armstrong's attempts to turn aside the simple and powerful objection above are brave but ineffective. He is attempting the impossible: he wants to preserve the asymmetry of instantiation while identifying instantiation with a relation that is plainly symmetrical.

Surely he is committed to the asymmetry of instantiation by his rejection of bundle theories and his espousal of the "subject-attribute view." (Armstrong 2004, 140) I would go so far as to say that 'asymmetrical instantion' is a pleonastic expression: instantiation by definition is asymmetrical. After all, it is a technical term like 'compresence,' which is symmetrical by definition.

If instantiation falls, then so do atomic facts. For an atomic fact just is one or more particulars' instantiating a universal. Thin particulars, thick particulars, particulars-as-facts, and instantiation all go together. To reject one is to reject the others. But there is worse to come.

We now consider whether facts can survive supervenience upon their constituents. My thesis is that they cannot: the new view implies that there are no facts. In numerous passages Armstrong tells us that the existence of a and the existence of F-ness do not suffice for the existence of the fact of a's being F. That was his old view and it makes perfect sense. If Al is bald and Sal is fat, then Al exists and so does fatness; but it doesn't follow that Al is fat. This is because a fact is an item in addition to its constituents. This in turn is grounded in the contingency of the connection between particulars and universals. Not only are particulars and universals contingent, their connection is as well. That was the old view. But if instantiation is partial identity, then, while the particulars and universals remain contingent, the connection becomes necessary. Contingency, which used to reside within the 'guts' of each fact, no longer resides there.This implies that, given the actual particulars and the actual universals, each fact is necessary. (Armstrong 2004, 144) A fact cannot fail to have the very constituents it has, and the constituents it has cannot fail to form that very fact. Facts supervene on their constituents. But if facts supervene on their constituents, then there is no difference in point of existence beween the mereological sum a + F-ness and the fact of a's being F: both exist automatically given the existence of Al and fatness. There is a notional difference but no difference in reality and no possibility that a, which is F, might not have been F. As far as I can see, the new view implies that there simply are no facts. A fact that is not something in addition to its subfactual constituents is no fact at all.

That Armstrong's new view makes hash of his combinatorial theory of possibility should be obvious. That theory required the promiscuous combinability of thin particulars and universals. On the new view, however, there are no thin particulars, no first-order instantiation (strictly understood as asymmetrical), and no contingency within facts. Given the actual particulars and the actual universals, the only possible combinations are the actual ones. The actual world becomes one big block of necessity. If a is F, then a cannot (logically) fail to be F. How then accommodate the intuition that Al might not have been fat, that there is no logical or metaphysical necessity that he be fat? One could invoke counterparts in other worlds. Al in our world at time *t* is fat, but he has counteparts in other worlds who are not fat at *t*. But such a scheme does not comport with Armstrong's naturalism according to which reality is exhausted by the space-time system, *this* space-time system.

Finally, what becomes of truth-making on the new view? The best argument for facts is the Truth-Maker Argument sketched above. At least some truths need truth-makers. They require ontological grounds of their being true. In some cases, an Armstrongian 'blob' will do the trick: if it is true that *a* exists, then it would seems that the existence of *a* alone suffices to make the truth-bearer true. And the same goes if *a* is essentially F. If *a* cannot exist without being F, then it would seem that the mere existence of a would suffice as truth-maker. But it other cases we need an Armstrongian 'layer-cake': if it is contingently true that *a* is F, then neither *a* nor F-ness alone suffice as truth-makers. There is need of a proposition-like entity as truth-ground. Enter facts or states of affairs. It is the fact of *a*'s being F that grounds the truth of '*a* is F.' Here then we have an argument for facts as truth-makers (assuming that other truth-making candidates can be excluded). But if instantiation is partial identity, then, as Armstrong puts it, *a* is necessarily F: *a* cannot exist without being F. (And given the partial identity of particulars and universals, F-ness cannot exist without being instantiated by *a*!) But if *a* is necessarily F, then *a* alone suffices as truth-maker for '*a* is F' and the argument from truth-making to facts collapses.

6 Facts as abstract objects: Reinhardt Grossmann

There seem to be insurmountable problems with facts on the concretist conception, facts as truth-making denizens of space-time. So we turn to the abstractist conception of Reinhardt Grossmann. It fares no better in my judgment. But there is not the space to canvass all the arguments. I will present two. The first is an inconclusive Grossmannian argument against concrete facts; the second is an argument against Grossmann's abstractist conception. First, some preliminaries.

Like Armstrong, Grossmann maintains both that properties and relations are universals and that they cannot exist unexemplified. Unlike Armstrong, Grossmann maintains that universals are abstract where 'abstract' means 'not spatiotemporal.' (Grossmann 1992, 7) It follows that Grossmann's universals, unlike Armstrong's, cannot be ontological constituents of spatiotemporal particulars or constituents of space-time itself. If they were, they would either be spatiotemporally located or constitutive of spatiotemporal locations. As remarked earlier, immanency and constituency are distinct notions: first-order universals that cannot exist unexemplified, and are in this sense immanent as opposed to transcendent, need not be constituents of the particulars that exemplify them.

And while both philosophers agree that there are facts, Grossmann holds that they, like universals, are abstract, including those facts with concrete constituents. (Grossmann 1992, 73-84)The fact of *Al's being fat* (or as Grossmann would express it, the fact that Al is fat) has a concrete subject constituent, Al himself, and an abstract property constituent, but the fact itself is abstract. It is a hybrid entity. The concrete particular exemplifies the abstract universal where exemplification is a full-fledged relation, and itself an abstract entity. Exemplification in a case like this spans the chasm separating the concrete realm of time and change from the timeless realm of Platonica. And the same holds for every fact that involves concrete particulars, even those that do not include an exemplification relation. The fact that Al loves Beatrice, for example, does not on Grossmann's view include an exemplification relation, a point on which he draws fire from Armstrong. (Cumpa and Tegtmeier (eds.), 2009, 48-51) But it too unites the concrete and abstract realms. Facts are the fundamental ontological category for Grossmann precisely because they bring together the two realms that Plato had sundered. (Grossmann 1990, 129; Grossmann 1984, 114)

It follows that for Grossmann, ordinary concrete particulars are not facts. This is a key difference with Armstrong for whom ordinary (thick) particulars are facts. With considerable injustice to the historical Aristotle and the historical Plato, we can say that Armstrong's position is 'Aristotelian' in that he brings universals 'down to earth' from 'Plato's heaven' whereas Grossmann's position is Platonic in that he leaves universals 'in heaven' but connects them to the concrete particulars here below by means of the nexus of exemplification. His facts, then, bridge the gap between the particulars in the realm of time and change and the universals in the timeless realm of Forms. For this reason, Grossmann's particulars cannot be facts. And because concrete particulars are not facts, Grossmann does not view the properties of such particulars as constituents of them. Facts have ontological constituents, but concrete particulars are not facts. Why not?

6.1 The localization argument against concrete facts

Consider a white billiard ball, A. If A is a fact, then some fact has a size, shape, color, and location. "But facts, it seems to me, do not have shapes and sizes." (Grossmann 1992, 29) And if they don't have shapes and sizes, then they don't have colors and locations. We are invited to conclude that concrete spatiotemporal particulars are not facts. It is easy to see that this argument does not settle the question. It appears merely to beg it by assuming that facts are abstract, i.e., not spatiotemporal. For if ordinary particulars are facts, then some facts do have size, shape, etc.

A stronger consideration is that if billiard ball A is a fact, then, at the place where A is located, there is a whole. But this cannot be a spatial whole, says Grossmann, because the 'is' in 'A is white' does not pick out a spatial relation, but the relation of exemplification. (Grossmann 1992, 29) A billiard ball, however, is a spatial whole having spatial parts. So a billiard ball is not a fact. This argument is not decisive either. Why could not a whole of spatial parts also be a 'whole' of ontological 'parts'? If a billiard ball is fact, then it has both spatial parts and ontological constituents, with the relation of exemplification among the latter.

Grossmann sketches a third argument. On Armstrong's approach one distinguishes between the thick and the thin particular, where thick particulars are facts. The thick particular A factors into the thin particular *a* and its nature N-ness. N-ness is a conjunctive property each conjunct of which is an intrinsic property of A. So for Armstrong, A = *a's being N* where whiteness is included in the nature N-ness. But then what are we saying when we say that A is white? We are not talking about the thin particular, *a*. We are talking about the thick particular, A. But the thick particular is a fact. According to Grossmann, this is a mistake: we can't be saying that the fact of *a's being N* exemplifies whiteness. Talk about individual things is not talk about facts, and conversely. Individual things are colored, but no fact is colored. (29) But Armstrong has a response. "I suspect that what is needed is a translation of ordinary subject-predicate talk into fact talk, a move which does not look too difficult to make." (Cumpa, 38) The idea, I take it, is that 'This billiard ball is white' is translatable by 'This billiard ball has whiteness as one of its property constituents' or 'A's nature N-ness includes whiteness.'

As I see it, none of Grossmann's arguments decisively refutes Armstrong's conception. And we are about to see that Grossman's conception is open to serious objection.

6.2 The 'bare particular' objection to abstract facts

As Grossmann rightly maintains, if particulars are bare, then, unlike Aristotelian substances, they do not have natures or essences. (Grossmann 1974, 97) Equivalently, if particulars are bare, then their properties cannot be divided into essential and accidental. This amounts to saying that the connection between a particular and its properties is the external nexus of exemplification. What makes a particular bare is not its having no properties, but the manner in which it has the properties it has. On this understanding of 'bare particular,' which is Grossmann's own, his concrete spatiotemporal particulars are all of them bare. For all of their properties are 'outside' of them, and tied to them by the external relation of exemplification. There is nothing in the definition of 'bare particular' to require that such particulars be constituents of ordinary particulars as they are for Bergmann and Armstrong. For Grossmann, ordinary concrete particulars are bare. They exemplify properties, but these properties are abstract (non-spatiotemporal) entities. They are not constituents of the things that exemplify them. Grossmann's particulars have spatial and temporal parts, but no ontological constituents. They are not facts, but constituents of facts, all of which are abstract, even those with concrete constituents.

If particulars are bare, then, to borrow a phrase from Butchvarov, they are "ontologically distant" from their properties with unpalatable consequences. (Butchvarov 1986, 131) Suppose Max, a billiard ball, is white. Whiteness is an intrinsic property of Max: he is not white in virtue of a relation to something else. Being white is thus unlike the property of being 12 inches from Moritz, a second billiard ball. But if Max is a bare particular, then he is white in virtue of standing in the external relation of exemplification to the abstract object, whiteness. It follows that Max, in himself, is not white. He is no more intrinsically white, white in his own nature, than he is intrinsically 12 inches from Moritz. We are thus brought back to Quilter's Antinomy of the Bare Particular discussed above. Max is intrinsically white and Max is not intrinsically white. That Max is intrinsically white is a datum, and that he is not intrinsically white is a consequence of Grossmann's theory of property possession.

Armstrong found a way around the antinomy by construing ordinary particulars as facts or states of affairs with thin particulars and universals as their ontological constituents. By bringing thin particulars and universals together in thick particulars construed as truth-making facts, Armstrong dramatically lessened the "ontological distance" between them. But this escape route is not available to Grossmann. His facts are not truth-making concreta but chasm-spanning abstracta linking spatiotemporal – or in the case of thoughts merely temporal – particulars with non-spatiotemporal universals.

7 Concluding aporetic postscript

David Armstrong was well-known for his intellectual honesty, and we should all strive to imitate this virtue of his. What intellectual honesty demands, however, is not clear when we get down to cases. In the present case it demands of me a recognition of the force of the Truth-Maker Argument for concrete facts, but also a recognition of how problematic facts are. I don't know how to get past this impasse. I conjecture that some if not all of the perennial problems of philosophy are genuine but insoluble. But I haven't show this. Not by a long shot. It is unlikely that it could be shown to the satisfaction of all competent practitioners. But I may have contributed something to an appreciation of the difficulty of one set of problems in ontology.

Bibliography

Armstrong, D. M. 1978. *Universals and Scientific Realism, Vol. 1, Nominalism and Realism*. Cambridge: Cambridge University Press.
Armstrong, D. M. 1983. *What is a Law of Nature?* Cambridge: Cambridge University Press.
Armstrong, D. M. 1989a. *Universals: An Opinionated Introduction*. Boulder: Westview Press.
Armstrong, D. M. 1989b. *A Combinatorial Theory of Possibility*. Cambridge: Cambridge University Press.
Armstrong, D. M. 1993. "A World of States of Affairs". *Philosophical Perspectives*, 7, 429–440.
Armstrong, D. M. 2004a. "How Do Particulars Stand to Universals?" In Zimmerman, D. (ed.), *Oxford Studies in Metaphysics, vol. 1*. Oxford: Oxford University Press, 139–154.
Armstrong, D. M. 2004b. *Truth and Truthmakers*. Cambridge: Cambridge University Press.
Armstrong, D. M. 2009. "Questions about States of Affairs". In Reicher, M. E. (ed.), *States of Affairs*. Frankfurt: Ontos Verlag, 39–50.
Armstrong, D. M. 2010a. *Sketch for a Systematic Metaphysics*. Oxford: Oxford University Press.
Armstrong, D. M. 2010b. "Reinhardt Grossmann's Ontology". In Cumpa, J. (ed.), *Studies in the Ontology of Reinhardt Grossmann*. Frankfurt: Ontos Verlag, 29–43.
Baxter, D. 2001. "Instantiation as Partial Identity". *Australasian Journal of Philosophy*, 79(4), 449–64.
Bergmann, G. 1967. *Realism: A Critique of Brentano and Meinong*. Madison: The University of Wisconsin Press.
Butchvarov, P. 1979. *Being Qua Being: A Theory of Identity, Existence, and Predication*. Bloomington: Indiana University Press.
Butchvarov, P. 1986. "States of Affairs". In Bogdan, R. (ed.), *Roderick M. Chisholm*. Dordrecht: D. Reidel, 113–133.
Butchvarov, P. 2010. "Facts". In Cumpa, J. (ed.), *Studies in the Ontology of Reinhardt Grossmann*. Frankfurt: Ontos Verlag, 71–93.
Chisholm, R. 1976. *Person and Object: A Metaphysical Study*. La Salle: Open Court.
Cumpa, J. and Tegtmeier, E. (eds.), 2009. *Phenomenological Realism Versus Scientific Realism: Reinhardt Grossmann-David M. Armstrong Metaphysical Correspondence*. Frankfurt: Ontos Verlag.
Frege, G. 1960."On Concept and Object". In Geach P. and Black, M. (eds.), *Translations from the Philosophical Writings of Gottlob Frege*. Oxford: Basil Blackwell, 42–55.
Frege, G. 1976. "Der Gedanke". In Patzig, G. (ed.), *Logische Untersuchungen*. Goettingen: Vandenhoeck and Ruprecht, 30–53.
Grossmann, R. 1974. "Bergmann's Ontology and the Principle of Acquintance". In Gram, M. S. and Klemke, E. D. (eds.), *The Ontological Turn: Studies in the Philosophy of Gustav Bergmann*. Iowa City: University of Iowa Press, 89–113.

Grossmann, R. 1983. *The Categorial Structure of the World*. Bloomington: Indiana University Press.

Grossmann, R. 1984. *Phenomenology and Existentialism: An Introduction*. London: Routledge and Kegan Paul.

Grossmann, R. 1990. *The Fourth Way: A Theory of Knowledge*. Bloomington: Indiana University Press.

Grossmann, R. 1992. *The Existence of the World: An Introduction to Ontology*. London: Routledge.

Mulligan, K., Simons, P. and Smith, B. 2009. "Truth-makers". In Lowe, E. J. and Rami, A., *Truth and Truth-Making*. Montreal: McGill-Queen's University Press, 59–86.

Mumford, S. 2007. *David Armstrong*. Montreal: McGill-Queen's University Press.

Plantinga, A. 1974. *The Nature of Necessity*. Oxford: Oxford University Press.

Strawson, P. F. 1950. "Truth". *Aristotelian Society*, Suplementary Volume 24, 136–137.

Vallicella, W. F. 2000. "From Facts to God: An Onto-Cosmological Argument". *International Journal for the Philosophy of Religion*, 48, 157–181.

Vallicella, W. F. 2002. *A Paradigm Theory of Existence: Onto-Theology Vindicated*. Dordrecht: Kluwer Academic Publishers.

Javier Cumpa
Armstrong's Hidden Substantialism

1 Introduction: Is Factualism a Truth of Armstrong's Ontology?

Factualism is the categorial thesis of Wittgenstein and Russell according to which the world is a world of states of affairs. This thesis stands in opposition to Substantialism, namely, the Aristotelian thesis that the fundamental category of the world is the category of substance or particular.

From *Universals and Scientific Realism* to *A World of States of Affairs* and *Truth and Truthmakers*, Armstrong has defended the hypothesis that the world is a world of states of affairs:

> We come now to the Factualist thesis that the world is a world of states of affairs. It is fundamental to the methodology of the present inquiry that Factualism is put forward here as a hypothesis only, and a philosopher's hypothesis at that. (Armstrong 1997, 8)

Why Factualism?[1] Armstrong has appealed to states of affairs to solve two pressing problems:

(P1) *The problem of the multiple location of properties*: what in the world makes properties capable of multiple location? (Armstrong 1989, 98).

(P2) *The problem of the ontological ground of contingent truths*: what in the world makes contingent truths true? (Armstrong 1997, 115).

In this paper I argue first that Armstrong's world is Substantialistic because he categorizes states of affairs as particulars. Secondly I show two unwelcome consequences of Armstrong's hidden Substantialism. These two consequences are that, since the Factualist statement that

(F) the world is a world of states of affairs (Armstrong 1997, 1)

is not ultimately a truth of Armstrong's ontology, his solutions to (P1) and (P2) cannot be true either.

---

[1] I have defended that the categorial fundamentality of states of affairs resides in their *descriptive* and *explanatory* powers. For more details, see Cumpa 2011 and 2014a.

Javier Cumpa: University of Miami, email: javiercumpa@miami.edu

2 States of affairs and the problem of universals

In this section, I shall explore the relation between states of affairs and Armstrong's solution to (P1).

Consider two red flowers:

(1)   this flower is red

and

(2)   that flower is red.

Armstrong tries to solve (P1) by claiming that the property of being red can be said to be capable of being 'in' (1) and (2) at the same time if and only if this 'in' is the 'in' of constituents of states of affairs:

> If two things have the very same property, then that property is, in some sense, "in" each of them. (Armstrong 1978, 108)

And he adds:

> Universals are constituents of states of affairs. (Armstrong 1989, 99).

This 'in,' as Armstrong points out, is the so-called "fundamental tie" of his theory of properties:

> It is often convenient to talk about instantiation, but states of affairs come first. If this is a "fundamental tie", required by relations as much as by properties, then so be it.(Armstrong 1997, 118)

Now, why this 'in' of states of affairs allows for the property of being red to be at the same time 'in' (1) and (2)? Armstrong's answer here seems to be that the 'in' of states of affairs, in contrast, for instance, with the 'in' (*part-whole relation*) of mereological sums, is not mereological at all. This nonmereological part-whole relation allows for the property of being red to be capable of being shared by (1) and (2):

> States of affairs hold their constituents together in a non-mereological form of composition, a form of composition that even allows the possibility of having different states of affairs with identical constituents. (Armstrong 1997, 118)

For Armstrong, if (1) and (2) had a mereological form of composition, these would supervene upon their parts. And if (1) and (2) were identical with their parts, no part of (1) could be also a part of (2). In other words, the property of being red could not be a common or shared part of (1) and (2). Armstrong establishes a connection between mereological composition and Moderate Nominalism (*Trope Theory*), i.e., the view that properties are not capable of being shared:

> the rules of composition for possible states of affairs that involve only tropes are somewhat nearer to the rules for whole and part. (For instance, if R is a nonsymmetrical relation, a Universals theory has the possibility of two wholly distinct states of affairs: *aRb* and *bRa* composed of the very same constituents. With tropes the two Rs could not be identical. (Armstrong 1989, 111)

The same strong link seems to be also established by Armstrong between non-mereological composition and Realism (*Immanent Realism*), the view that properties are capable of being shared:

> the complete constituents of a state of affairs are capable of being, and may actually even be, the complete constituents of a different state of affairs. Hence constituents do not stand to states of affairs as parts to whole. (Armstrong 1989: 92)

In this regard, we can perfectly describe Armstrong's solution to (P1) in the following terms:

(S1)   Properties are capable of multiple location in virtue of the 'in' of states of affairs.[2]

3 States of affairs and the problem of truth

Now we will look at the relation between states of affairs and Armstrong's solution to (P2). What of the ontological ground of contingent truths? Let's consider a contingent truth such as

(3)   This apple is red

According to Armstrong, the *mereological sum* formed by the apple, the property of being red, and instantiation cannot serve as truthmaker for (3). As Armstrong notes, the parts of this particular whole do not stand in relations to each other. So, the apple and the property of being red could be parts of the very same mereological sum even standing in no relation whatsoever among them:

---

[2] I show the logical inconsistency of Armstrong's ontology of instantiation in Cumpa 2014b.

> Why do we need to recognize states of affairs? Why not recognize simply particulars, universals (divided into properties and relations), and, perhaps, instantiation? The answer appears by considering the following point. If a is F, then it is entailed that a exists and that the universal F exists. However, a could exist, and F could exist, and yet it fail to be the case that a is F (F is instantiated, but instantiated elsewhere only). a's being F involves something more than a and F. It is no good simply adding the fundamental tie or nexus of instantiation to the sum of a and F. The existence of a, of instantiation, and of F does not amount to a's being F. (Armstrong 1989, 88)

Thus Armstrong concludes that it must be not the sum, but something else, namely, the state of affairs

(4) This apple is red

that, in bringing together the apple and the property of being red, makes (3) true.[3] As Armstrong rightly puts it:

> We are asking what in the world will ensure, make true, underlie, serve as the ontological ground for, the truth that a is F. The obvious candidate seems to be the state of affairs of a's being F. In this state of affairs (fact, circumstance) a and F are brought together. (Amstrong 1997, 116)

In this sense, we can describe Armstrong's solution to (P2) as follows:

(S2) states of affairs make contingent truths true.

It seems clear to me that in Armstrong's ontology the truths of (S1) and (S2) depend upon the truth of (F).

4 The categorial clash between factualism and the victory of particularity

I shall argue here that (F) cannot be a truth of Armstrong's ontology because he categorizes states of affairs as particulars. In accordance with the so-called "vic-

---

[3] If states of affairs are *grounds of truth*, I fail to understand why Armstrong (1997, 10) characterizes Factualism (a) as a *mere hypothesis* and (b) as having *lower epistemic credit* than Physicalism and Naturalism.

tory of particularity," he claims that states of affairs are particulars.[4] On Armstrong's view, there is no category of state of affairs.

As Armstrong expresses it:

> States of affairs contain as constituents both particulars and universals. But what of the states of affairs themselves? Should they be classified as particulars, universals or neither? Confining ourselves here to first-order states of affairs, the only ones that have been so far considered, the answer would appear to be that they are particulars. For they lack the repeatability that is the special mark of universals. (…) First-order states of affairs are (first-order) particulars. This is the "victory of particularity". For first-order states of affairs, particulars + universals = a particular. (Armstrong 1997, 126)

If the entity *this apple is red* belongs to the category of the entity *this apple*, then (F) cannot be a truth of Armstrong's ontology. Quite the contrary. It is the Substantialist statement that

(S)   the world is a world of particulars

that should be regarded as a truth of Armstrong's ontology. So, Substantialism wins the grand final!

It is true that Armstrong (1997, 126) has stressed that the victory of particularity is no big deal for states of affairs. For atomic states of affairs are after all "thick" or "first-order" particulars, i. e., particulars having (non-relational) properties. However, it seems to me that there is a plain reason why the particulars of (S) could not be understood as thick or first-order particulars.

Let's concede that Armstrong is right: states of affairs are particulars. What we have to ask immediately is: Do states of affairs belong to the category of thin particular or to the category of thick particular? I think that it is pretty obvious that the category of particular in question is the category of thin particular. The reason is that Armstrong claims that states of affairs belong to the same category as one of their constituents, the non-repeatable thin particular.

---

[4] In Cumpa 2012, I argue that Armstrong's argument for the victory of particularity is grounded in a *category-mistake*.

5 Concluding remarks: the ontological consequences of the clash

Since (S) is a truth of Armstrong's ontology, it is clear that (S1) and (S2) cannot be true.

If my discussion of Armstrong is on the right track, he would seem to have to accept that

(5)   Properties are parts of thin particulars

and

(6)   Truthmakers for contingent truths are thin particulars.

To sum up, the moral of this metaphysical story is that if Factualism is not a truth of Armstrong's ontology, then he can solve neither (P1) nor (P2). As I see it, these consequences show Armstrong's need for recognizing the category of states of affairs in his ontology.

Acknowledgments: I would like to thank Otávio Bueno, Francesco Calemi, Anthony Fisher, and Erwin Tegtmeier for helpful comments.

Bibliography

Armstrong, D. M. 1978a. *Universals and Scientific Realism, Vol. 1, Nominalism and Realism*. Cambridge: Cambridge University Press.
Armstrong, D. M. 1989. *Universals: An Opinionated Introduction*. Boulder: Westview Press.
Armstrong, D. M. 1997. *A World of States of Affairs*. Cambridge: University Press.
Armstrong, D. M. 2004. *Truth and Truthmakers*. Cambridge: Cambridge University Press.
Cumpa, J. 2011. "Categoriality: Three Disputes over the Structure of the World". In Cumpa, J. & Tegtmeier, E. (eds.), *Ontological Categories*. Frankfurt: Ontos Verlag, 15–65.
Cumpa, J. 2012. "Observation and Interpretation. The Problem of the Problem of Universals". *Metaphysica*, 13(2), 131–143.
Cumpa, J. 2014a. "A Materialist Criterion of Fundamentality". *American Philosophical Quarterly*, 51(4), 319–324.
Cumpa, J. 2014b. "Exemplification as Molecular Function". *Philosophical Studies*, 170(2), 335–342.

Kristie Miller
Persisting Particulars and their Properties

1 Introduction

> My present inclination is to say that both identity and relational analyses are intelligible hypotheses. I reject the identity analysis, looking rather to relations between different phases to secure the unity of a particular over time. But I do not think that the identity view can be rejected as illogical. If it is to be rejected, then I think it must be rejected for Occamist reasons. The different phases exist, and so do their relations. These phases so related, it seems, are sufficient to secure identity through time for all particulars. I suggest, then, that the identity view of identity through time is not illogical. The question is rather whether it is a postulation which is fruitful, or expedient, or which we are compelled, to make. (Armstrong 1980, 70)

What Armstrong, above, calls the relational analysis is what is now more commonly known as perdurantism. It is the view that objects persist through time by being composed of numerically distinct particulars (instantaneous temporal parts) each of which exists at (and only at) a particular instant, and by the obtaining, between said distinct particulars, of certain relations. These relations are typically held to be (as Armstrong held them to be) similarity cum causal relations. Thus persisting objects are four-dimensional – which is why perdurantism is sometimes also known as four-dimensionalism: they are only ever partially present at any one time in virtue of having some part present at that time, while the whole persisting object occupies a four-dimensional region.[1] By contrast, what Armstrong calls the identity analysis is what is now more commonly known as endurantism. It is the view that objects persist through time by being wholly present at different times. If x endures from t to t* then there is something that is wholly present at t, and something that is wholly present at t*, and the thing wholly present at t is identical to the thing wholly present at t*. Indeed, endurantism is also sometimes known as three-dimensionalism since it is taken to be the view that persisting objects are three-dimensional. Any persisting thing occupies a se-

---

[1] Defenders of this view include Balashov 1999; 2000a; 2000b; 2000c; 2002; 2003, Heller 1984; Lewis 1983; Sider 2001; Hawley 2001.

Kristie Miller: University of Sydeny, email: kristie_miller@yahoo.com

ries of three-dimensional regions; but what occupies each such region is one and the same thing.²

In his (1980) Armstrong argues, on Occamist grounds, that perdurantism is the preferable view. He does so by attempting to determine what sort of ancillary metaphysical claims one would need to embrace in order to make sense of the endurantist claim that, as he puts it, the phases of a persisting object are numerically identical. Armstrong thinks there are ancillary metaphysical claims that allow us to make sense of the endurantist view, which is why he takes endurantism to be a coherent view. But he finds none of these metaphysical claims attractive. Moreover, he notes, we do not seem to need to posit an identity relation between the phases of persisting objects to explain how it is that objects persist. Thus, he concludes, we ought to prefer perdurantism to endurantism.

By and large this paper is directed neither at disputing the conclusion that we ought to prefer perdurantism to endurantism, nor to defending it, though at the very end of the paper this issue will be reconsidered. In the main it takes issue with Armstrong's reasons for arriving at his conclusion. In essence Armstrong attempts to spell out, on behalf of the endurantist, potential answers to the question: in virtue of what is one phase of an object numerically identical to some other phase of the object. In what should the endurantist suppose that identity over time consists? He then suggests that any answers to this question are either independently implausible, or else, in the best case scenario, they commit the endurantist to a metaphysical picture to which the perdurantist need not be committed. Thus, the thought is, in order to secure the relevant identity through time the endurantist needs to take on additional metaphysical machinery of some kind or other; machinery that may well be implausible and machinery which the perdurantist does not need. This paper will argue, *inter alia*, that that contention is false.

This paper takes as its starting point some of the suggestions Armstrong canvasses on behalf of the endurantist and builds on these in the light of more recent literature on persistence. First, however, the paper argues that it is a desideratum of any metaphysical account of persistence that it allow us to distinguish, by appealing only to features of that account, competitor views about the manner in which objects persist. Call this desideratum DISTIGUISHABILITY. According to DISTINGUISHABILITY, any adequate way of understanding endurantism as a theory of persistence must allow us to distinguish it from its competitor theories. Since Armstrong wrote his (1980), however, the number of competitor theories of persistence has grown. In this paper it will be argued, in sections 3 and 4, that

---

2 Defenders of this view include van Inwagen 1987; 1990; Thomson³ 1983; 1998; Merricks 1994; 1999; Lowe 1987; Haslanger 1989;

the suggestions Armstrong offers on the behalf of the endurantist fail to meet DISTIGUISHABILITY. In section 5 I move on to consider an alternative characterisation of endurantism that does meet DISTINGUISHABILITY. In that section it is argued that given this characterisation the endurantist has the same options open to her as does the perdurantist with respect to the various ancillary metaphysical theses that Armstrong considers. Thus there is no straightforward Occamist argument in favour of perdurantism over endurantism.[4] In the final section I will, however, gesture towards some explanatory considerations that might push the endurantist towards accepting ancillary metaphysical commitments that the perdurantist need not accept and thus might ultimately vindicate an Occamist style argument.

2 Transdurantism

Though endurantism and perdurantism are the leading views about the manner in which objects persist through time there is a recent competitor view. In Miller (2009) I called this view terdurantism, and since then Daniels (2014) has called it transdurantism. This is the view according to which persisting objects are four-dimensional objects that lack (temporal) parts. In essence such objects, though extended along the temporal dimension, are *extended temporal simples* since at each time at which they exist they do so without there existing a proper part of them that exists at that time and overlaps every spatial part of them at that time. If such objects also lack spatial parts then they are four-dimensionally extended simples. In what follows, however, I will assume that transduring objects typically do have spatial parts: they merely lack temporal parts.

On the one hand, like perdurantism, transdurantism holds that persisting objects are four-dimensional not three-dimensional. Indeed, there is nothing, according to the transdurantist, that is wholly present at each three-dimensional region occupied by a transduring object.[5] A transduring object is, in a sense to be explicated, wholly present at a four-dimensional region. While the perdurantist

---

[4] Indeed this must be the case if, as I have argued elsewhere, endurantism and perdurantism are metaphysically equivalent theories. See Miller 2005.
[5] At least, that seems intuitive to be the case in the same way that no extended simply is wholly present at any of its locations. Whether that is so will depend on exactly how one defines the notion of being wholly present. Parsons 2007 for instance, defines it in such a way that transduring objects are wholly present at each moment they exist and Daniels 2014 appeals to this terminology. Since I do not find this a helpful way to think of such objects I will define the notion of being wholly present in a way that does not have this consequence.

agrees that persisting objects are four-dimensional she thinks this is so because such objects are composed of further objects, temporal parts, and that each instantaneous temporal part is wholly present at a three-dimensional region. Thus the perdurantist, like the endurantist but unlike the transdurantist, thinks that there is something wholly present at each three-dimensional region: the two merely disagree about what that thing is (a temporal part of the perduring whole, or the whole object). Thus, unlike the endurantist, the transdurantist does not think that a persisting object is wholly present at a series of different times; instead, she holds that persisting objects are spread out through time but are spread out without having (temporal) parts.

I have just offered a rough characterisation of the three competitor views in order simply to give the reader a sense of each. No attempt was made to define the views more precisely since the remainder of the paper is devoted to thinking about how to distinguish these views.

As Armstrong (and many others after him) have seen it, the endurantist needs to explain what it means, and how it can be, that an enduring object is wholly present across time and numerically identical at each of those times. Indeed, in the light of the previous rough characterisations of the three views we need to be able to define endurantism in such a way that it is distinguishable from its competitors. I call this desideratum DISTINGUISHABILITY:

DISTINGUISHABILITY: Whatever resources we need to define endurantism should be such that, by appealing only to those resources, we can distinguish endurantism from all of its competitors.

In what follows I begin by outlining some of the proposals Armstrong develops on behalf of the endurantist; these are proposals that, as he sees it, allow the endurantist to explain what it means to say that objects endure through time. As such we can parlay these resources into resources for defining endurance. As I will subsequently argue, however, while these might succeed in the context in which Armstrong found himself – that of explicating endurantism in contrast to perdurantism – they do not allow the endurantist to distinguish her view from transdurantism and thus they fall foul of DISTINGUISHABILITY.

3 Objects as property bundles

Let us momentarily set aside transdurantism and consider matters from the perspective from which Armstrong in his (1980) considered them. The question arises, how can a thing at one time be the selfsame thing as a thing at some other time, particularly if the former manifests different properties to the latter? One answer lies in thinking about the relationship between properties and particulars. According to one such view particulars just are bundles of properties. In what follows I suppose that the locution 'bundle' is doing significant work here, and that it is with respect to this notion that accounts disagree. That is, I assume that these views hold that objects are identical to bundles, and the issue is then what relationship bundles bear to pluralities of properties: being composed of them, supervening on them, being grounded by them and so forth. Nothing in what follows will be sensitive to such differences.

Suppose that objects are bundles of properties. Which properties? There are at least two options. On the first, particulars are bundles of qualitative properties. On the second, particulars are bundles of properties that include a non-qualitative property: a thisness or haecceity.

One possibility is that enduring objects are objects that persist over time and, at each time, share the same thisness.[6] One way to understand that proposal is in terms of the second option, above. The thought is that it is sameness of thisness over time that constitutes sameness of identity over time. Why introduce a thisness? Well the following three claims all seem plausible and yet are apparently inconsistent:

1. Persisting objects change over time
2. Leibniz's Law is true
3. Persisting objects endure.

To see why these three claims are apparently inconsistent suppose that a persisting object, O, is red at one time (t1) and blue at another (t2). Then it seems as though O has different properties at t1 than it does at t2, and thus that whatever exists at t1 is not, given the truth of Leibniz's Law, numerically identical to the thing that exists at t2. That, of course, is exactly what perdurantists say: they take it that at t1 there exists a three-dimensional object, P1, and at t2 there exists a distinct three-dimensional object, P2, and that P1 and P2 are each temporal parts of O. One strategy the endurantist might adopt for responding to this apparent inconsistency would be to reject (2). One could then reconcile change over time with endurance by holding that at each moment an enduring object exists there exists

---

[6] This is not a proposal that Armstrong considers, though he does consider similar (though importantly different) proposals (as we shall see in the following section) according to which identity consists in sameness of substratum.

a distinct bundle of properties – distinct insofar as it is composed of somewhat different qualitative properties – but such that each bundle includes the same non-qualitative thisness. Thus an enduring object changes over time by instantiating different qualitative properties at different times, but is one and the same object at each time it exists since it instantiates the same non-qualitative thisness at each such time. This account will, however, only work if we reject (2) since on such an account the properties an enduring object instantiates at different times are different despite it being one and the same object at each such time, in contravention of Leibniz's Law.

According to this proposal we would define endurance in something like the following manner:

> END (1): An object endures through some temporal interval, T iff (i) for any two times, t and t*, in T, a bundle of properties, B, exists at t and a bundle of properties B* exists at t* and (iii) non-qualitative thisness T* is part of bundle B and non-qualitative thisness T** is part to of bundle B*.

Arguably, this definition of endurance distinguishes endurance from perdurance since perdurantists will almost certainly hold that if O perdures then it is not the case that every bundle that exists at each time at which O exists (and which is O at each of those times) has amongst its properties the very same non-qualitative thisness. Even so, however, the definition nevertheless falls foul of DISTINGUISHABILITY. For it does not distinguish endurance from transdurance. After all, transduring objects might instantiate the same non-qualitative thisness at each time at which they exist: indeed, it might be in virtue of instantiating the same thisness at each time, that a transduring object is the same (albeit four-dimensional) object across time.

Thus even if the endurantist is inclined to posit non-qualitative thisnesses and to give up Leibniz's Law, doing so does not, in itself, help her to distinguish endurance from transdurance. She will need additional resources. I turn to these in the following section and argue that once she has these additional resources she has no need to appeal to non-qualitative thisnesses.

Contemporary endurantists, however, typically do not want to jettison Leibniz's Law. So the strategy just outlined will be unattractive to them. Instead, contemporary endurantists attempt to preserve all of (1) to (3) above. They do so by temporally relativising the having of properties. They hold that enduring objects do not instantiate properties *simpliciter*. Rather, properties are disguised relations to time. Or, alternatively, the having of properties is temporally relativised. On the former proposal O, which manifests redness at t1 and blueness at t2, has the temporally indexed properties of being red-at-t1 and blue-at-t2 (van Inwagen 1990).

On the latter proposal O has the temporally adverbialised properties of being red t1ly and blue t2ly (Haslanger 1989). The idea is that this resolves the tension between (1) to (3) because persisting object O is, at all times at which it exists, red-at-t1 and blue-at-t2 (or red t1ly and blue t2ly). Thus it instantiates the same temporally indexed properties at each time and there is therefore no inconsistency with Leibniz's Law. Nevertheless, it *manifests* different properties and thus appears to change because although it instantiates the very same set of temporally indexed properties at each time it exists, which property it manifests is sensitive to which time it is. At t1 it manifests redness not blueness, since it instantiates red-at-t1. Thus persisting objects do change and (1) is preserved as is (3).

Translated into talk of property bundles this is the view that an enduring object is identical to the very same bundle of properties at each time at which it exists. But if that is right then there is, at least *prima facie*, no work for a non-qualitative thisness to do. In answer to the question: in virtue of what is this three-dimensional thing, at t1, identical to that three-dimensional thing, at t2, the answer is: because the thing at t1 is a bundle of properties, P1, and the thing at t2 is a bundle of properties, P2, and P1=P2.

So let us suppose that the endurantist jettisons any appeal to non-qualitative thisnesses and holds that enduring objects are bundles of purely qualitative properties. For purely illustrative purposes let us suppose that there is an enduring object, O, that exists at only two times, t1 and t2, and has only two properties – it is red at t1 and blue at t2. Then, on the assumption that the endurantist does not want to reject Leibniz's Law, she will identify O with the following bundle of temporally relativised properties <red-at-t1, blue-at-t2>.[7] Then we might be tempted to define endurance as follows:

END (2): An object, O endures iff (i) O exists through some temporal interval T and (ii) for any two times, t and t* in T, 'O' at t picks out bundle B1, and 'O' at t* picks out bundle B2 and B1=B2.

But END (2) clearly will not do. It does not even distinguish endurance from perdurance. For it could be that O is a four-dimensional perduring object. If 'O' picks out a four-dimensional object then that four-dimensional object, at t, is identical to a bundle – the bundle of properties that is identical to four-dimensional O – and O at t* is also identical to that self-same bundle. As long as we can pick out O at different times by talking of O at t and O at t*, and we are still picking out the

---

[7] See for instance Benovsky 2006; 2008; 2010 for extensive discussion of this issue.

whole four-dimensional object than END (2) is consistent with O perduring.[8] Nevertheless, we can amend END (2) in light of the fact that the endurantist who does not want to jettison Leibniz's Law must think that the properties instantiated by persisting objects at times are irreducibly temporally relativised (in some manner or other). Thus we can say:

> END (3): An object, O, endures iff (i) O exists through some temporal interval T and (ii) for any two times, t and t* in T, 'O' picks out bundle B at t, and bundle B* at t* and (iii) the properties in bundle B and B* are irreducibly temporally relativised and (iv) B = B*.

Since the perdurantist rejects the claim that properties are irreducibly temporally relativised it is clear that this definition of endurance allows us to distinguish endurance from perdurance. Nevertheless, it, like END (1) fails to distinguish endurance from transdurance. This is because like the endurantist, the transdurantist must hold that properties are irreducibly temporally relativised. To see why, consider a simple, S, that is extended in space (but not time).[9]

Suppose S is variegated – it has different properties at different locations. But the simple has no proper parts at those locations that can instantiate properties simpliciter. The most straightforward response is to suppose that the simple instantiates irreducibly spatially relativised properties. That is, the simple has property p1-at-S1 and p2-at-S2 where S1 and S2 are spatial locations occupied by the simple. This strategy is strictly analogous to the one the endurantist endorses but instead of relativising properties to times, it relativises them to spatial locations.

Since transduring objects lack temporal parts but (we are assuming) possess spatial parts it is natural for the transdurantist to say that if a transduring object is cross-temporally variegated (i.e. if it changes over time) then since this cannot be because it has parts with those properties *simpliciter* it must be because the whole four-dimensional transduring object has irreducibly temporally relativised properties. But if so transduring objects will meet the conditions laid out in END (3).

So if END (3) is a viable definition of endurance then it would need to be argued that transdurantist need not, and indeed should not, be committed to irreducibly temporally relativised properties. One way to make that case would be to suggest that transdurantists should instead appeal to distributional properties.

---

[8] For more on the flexible semantics for perdurantists see Braddon-Mitchell and Miller 2006.
[9] For a defence of the idea of extended simples see Simons 2004; Parsons 2004 and Braddon-Mitchell and Miller 2006b.

Distributional properties are, roughly, properties that are extended (along some dimension or other). A typical example of a distributional property is the property of being polka dotted.[10] Here is an intuition one might have: there is something importantly different in the way in which enduring and transduring objects instantiate distributional properties. Let us say that temporal distributional properties are properties that are distributed across times. In contrast, spatially distributional properties are properties that are distributed across space. Spatio-temporally distributed properties are those that are distributed across both space and time. Then temporally indexed properties – at least, those indexed to temporal instants – will be spatially but not temporally distributed. By contrast properties such as ageing through some interval will be temporally distributed (and typically also spatio-temporally distributed).

Then, one might think, enduring objects are objects whose temporally distributed properties reduce to, or are grounded by, spatially distributed properties while transduring objects are objects whose spatially distributed properties reduce to, or are grounded by, either temporally distributed properties or spatio-temporally distributed properties.

Consider an object, O^*, that is green at t1, white at t2, green at t3, and white at t4. The endurantist thinks that O^* endures, the transdurantist thinks it transdures. Both agree that O^* instantiates the spatially distributed properties of being green-at-t1, white-at-t2, green-at-t3 and white-at-t4. Both agree that O^* has the spatio-temporally distributed property of being striped. It seems natural for the endurantist to think that O^* has the spatio-temporally distributed properties it does (such as being striped) in virtue of having the spatially distributed properties it does. After all, the endurantist thinks that O^* is wholly present at each moment it exists. The reason it has the spatio-temporally distributed property of being stripy is because at one time it is white, and another time green, and the next time white, and so forth. On the other hand, the transdurantist denies that O^* is wholly present at any moment. Instead she thinks O^* is spread out across a four-dimensional region. So while there is something special about the three-dimensional regions that O^*, qua enduring object, occupies (since it is wholly present at each such region) there is nothing special about the three-dimensional regions that O^*, qua transduring object, occupies. Instead, it is the four-dimensional region that is special given transdurantism. Given this, it makes sense for a transdurantist to suppose that transduring objects instantiate spatially distributed properties in virtue of instantiating spatio-temporally distributed properties. Thus O^*, qua transduring object, is green at one time and

---

10 For a defence of the need for distributional properties see Parsons 2004.

white at another – it is green-at-t1 and white-at-t2 – in virtue of instantiating a spatio-temporally distributed property – the property of being striped in a particular way. The most natural way to read the 'in virtue of' claims here is in terms of grounding. Without making any assumption about what grounding is (a relation between facts or an operator on sentences) anyone who thinks that there is grounding thinks that the world is, in some good sense, structured into the more and the less fundamental with the more fundamental grounding the less fundamental.[11] With this in mind there are two competing grounding theses on offer:

(1) GROUND: Spatially distributed properties are grounded by spatio-temporally distributed properties.
(2) GROUND: Spatio-temporally distributed properties are grounded by spatially distributed properties.

The endurantist will reject (1) and accept (2) while the transdurantist will accept (1) and reject (2). Thus the transdurantist and endurantist agree that O* is to be identified with the bundle of properties <green-at-t1, white-at-t2, green-at-t3, white-at-t4, Stripy@L> (where Stripy@L is the name of the particular spatio-temporally distributed property of being stripy). The endurantist, however, holds that the grounding structure of O* is <green-at-t1, white-at-t2, green-at-t3, white-at-t4> grounds <Stripy@L > while the transdurantist contends that the structure of O* is <Stripy@L > grounds <green-at-t1, white-at-t2, green-at-t3, white-at-t4>.[12]

The endurantist can distinguish endurance from transdurance if she is prepared to countenance grounding relations. To do so she would need to amend END (3) in something like the following manner:

END (4): An object, O, endures iff (i) O exists through some temporal interval T and (ii) for any two times, t and t* in T 'O' picks out bundle B at t and bundle B* at t* and (iii) the properties in bundle B and B* are irreducibly temporally relativised and (iv) B = B* and (v) spatio-temporally distributed properties are grounded by spatially distributed properties.

---

[11] See for instance Schaffer 2009 Trogdon (forthcoming) and Audi (forthcoming) for further discussion of the notion of grounding.
[12] It is worth noting that distinguishing endurantist and perdurantism in this manner does not prejudge the issue as to whether, in general, distributional properties are grounded by, or reducible to, non-distributional properties. Parsons 200X provides reasons to suppose that distributional properties are not reducible to non-distributional properties. For present purposes that is irrelevant since we are only interested grounding relations between distributional properties.

But if the endurantist needs to appeal to relations of grounding in order to distinguish her view from transdurantism then, it seems, Armstrong is right to suggest that Occamist considerations militate against her view. In section 4 I consider an alternative way to define endurance. First, however, I want to investigate a second option that Armstrong outlines: the view according to which objects are an amalgam of a substratum with properties and relations.

4 Objects as substrata with properties

According to what we can call *substratum theories*, particulars are a combination of a bare particular and properties. If the endurantist accepts this picture then, Armstrong suggests (1980, 70-72) she can hold that the identity of an object over time is grounded by the identity over time of a bare particular. The idea, I take it, is that the thing that occupies a particular three-dimensional region and the thing that occupies some other three-dimensional region is one and the same thing because the same bare particular occupies each region. That bare particular might have attached to it, at each region, different properties but those different properties are attached to one and the same substratum.

Armstrong attributes this view to Locke. The view is structurally similar in various ways to the view that enduring objects are bundles of qualitative properties combined with a thisness. As such both views face similar issues. Like its analogue bundle-theory view, this view will need to discard Leibniz's Law since according to it enduring objects that change instantiate different properties at different times. Again, analogously, this view can save Leibniz's Law if it combines the claim that the substratum remains the same over time, with the claim that properties are irreducibly temporally relativised, and that an enduring object instantiates, at all times, the same set of relativised properties. But, once again, it is hard to see that the endurantist, qua endurantist, has any need to appeal to a substratum to ground the identity of the enduring object over time if she holds that the enduring object instantiates, at every time, one and the set of irreducibly temporally relativised properties. There might be good reasons to posit substrata: but being an endurantist does not seem to be one of them.

Nevertheless, let us suppose the endurantist were inclined to reject Leibniz's Law and accept the view Armstrong attributes to Locke. Then, Armstrong argues, one is not out of the woods. The problem, as Armstrong sees it, (1980, 71) is that even if the smallest objects – the mereological simples – are like this, it is hard to make sense of how ordinary composite objects endure in this manner. Armstrong's worry is as follows. Consider some set of simples at one time, t, and the same set of simples at some other time, t*. It looks as though the thing the simples compose at t, and the thing the simples compose at t*, is a good candidate to be one and the same enduring object. Or, more carefully, it looks as though that object is a *better* candidate to be an enduring object that an ordinary object which gains and loses simples across time. I assume that Armstrong thinks so because he thinks it plausible that the substratum of a composite object at a time is some function of the substratum of each of the simples that compose that object at a time. If so, then if endurantism is true then mereological essentialism (the view that composite objects do not gain or lose parts) is also true. Then there are enduring objects, but no ordinary objects endure. If objects persist iff they endure, then ordinary objects do not persist. This is a view that Chisholm (1973; 1975; 1976) ultimately accepted for just these reasons.

Armstrong suggests that Locke, recognising these problems, advocates an identity theory of properties as well as substances (1980, 71-73). Armstrong's idea (attributed to Locke) is that in some cases the identity of an object over time lies in the identity, over time, of a substratum, and in other cases the identity of an object over time lies in the identity, over time, of a certain sort of property. Armstrong suggests, for instance, that one such property might be 'being a living thing'. Then if O! at t and O* at t* share that property then they are phases in the existence of the same living thing (O) even though there is (let us suppose) no substratum that is shared between O! at t and O* at t*.

As I read Armstrong, the proposal is that the grounds of identity are disjunctive and something like the following is the case:

END (5): An object, O endures iff O exists through some temporal interval, T, and for any instants t and t* in T either (a) the substratum of O at t is numerically identical to the substratum of O at t* or (b) O at t and O at t* share some property P.

If (a) obtains then O endures through some period in virtue of the existence of the same substratum through that time. If (b) obtains then O endures through some period in virtue of the identity over time of some property. It is a tantalising proposal that Armstrong presents on behalf of Locke. The problems, however, emerge quite quickly. Armstrong famously argues that properties are universals (Armstrong 1978). But, as Armstrong himself notes, if the sorts of properties that ground identity over time are universals then it will almost certainly be trivially true that almost any property is instantiated at more than one time by different substances. Consider the property of being a living thing. If that property is a universal then that very property is instantiated by a host of objects at different times that have a different substratum: it will be instantiated by every living thing at every time at which that thing is alive. If so, there can be no special identity set up between O's having the property of being alive at t and O*'s having the property of being alive at t*. After all, my dog has the property of being alive at t, and your giraffe has that property at t*, but we do not think, on this basis, that there is a single enduring thing that has a dog phase and a giraffe phase.

With this in mind Armstrong suggests (1980, 73) that Locke (or any endurantist wanting to appeal to this view) needs to think of properties as property instances not as universals or, as we might now say, Locke needs to appeal to tropes. Then the idea is that the, say, P-ness of O at t and the P-ness of O* at t* are themselves particulars and, moreover, that they are phases of the same property instance. That is, the P trope at t is numerically identical to the P trope at t*. Then we would define endurance as follows:

> END (6): An object, O endures iff (i) O exists through temporal interval T and for any two instants t and t* in T either (a) the substratum of O at t is numerically identical to the substratum of O at t* or (b) there is some trope, T, such that O at t has T and O at t* has T and T is the essence of O.

This proposal is a sort of half-way house between tying identity to a unique non-qualitative property (a thisness) and tying it to a universal qualitative property (being a living thing). Since tropes are particulars it side-steps the problem that it will be trivial for an x and y at two different times to share the same property instance: it will not be trivial. On the other hand, it opens up a plethora of other questions. After all, tropes are particulars like objects. So if we were puzzled by how an object can, at different times, be numerically identical, there seems no reason to be less puzzled by how a trope can manage the same feat. Indeed, the question now arises as to which tropes are such that they are, in effect, multi-located and ground identity over time and which are not.

I take it that the most perspicuous version of this view would hold that it is something like an essential property (but not a non-qualitative essence) that, in each case, grounds identity. Thus if I am person, and being a person is of my essence, then there is only one trope (or cluster of tropes) that are multi-located: those that constitute my being a person; and it is in virtue of the multi-location of this trope or tropes that I endure. What distinguishes this enduring person from that enduring person is that although they are both persons each has a distinct (multi-located) personhood trope.

The picture is a dual one with respect to identity. In some cases identity is grounded in sameness of substratum, and in other cases in sameness of some essential property instance. While on the face of it this might look a bit unappealing, the idea, I take it, is that enduring simples have their identity grounded in sameness of substratum while enduring composites have their identity grounded in the multi-location of an essential property instance.

Armstrong's concern about this picture is that it commits one to an ontology of tropes rather than universals, which he takes to be a substantial claim, and, he thinks, a false one. Even if he is wrong about the costs associated with trope theory, however, I think it would still represent some cost to the endurantist if she had to, as it were, hitch her wagon to a particular view about the nature of properties. If one supposes that a theory's being consistent with a wider array of views about the underlying ontology of the world is, other things being equal, a virtue of the theory over its competitors, then even in the best-case scenario this would give us some reason to prefer perdurantism to endurantism.

There are, however, problems with the view spelled out by Armstrong on behalf of Locke (and other endurantists). We have already noted that it requires jettisoning Leibniz's Law. I will say no more about this. The other problem is that END (6) does not allow us to distinguish endurantism from transdurantism and thus falls foul or DISTINGUISHABILITY.

Consider a case in which what grounds the identity of O is sameness of substratum over time. In what does this sameness consist? Does it consist in a substratum, S, being wholly present at t and a substratum S*, being wholly present at t*, and S = S*? Or is the substratum that exists at t, and the substratum that exists at t*, one and the same substratum in virtue of that substratum being temporally extended but temporally partless? If the former, then surely O endures; if the latter, then surely O transdures. Likewise, consider a case in which the identity of O* is grounded in the identity over time of a trope, T. In what sense is T identical over time? Is it that T is wholly present at t, and T*, is wholly present at t*, and T is numerically identical to T*? Or is it that T is temporally extended across t and t*, but without having any trope-parts at each location – that is, T exists at t and t* by being a temporally extended simple trope? If the former, then surely O* endures; if the latter, then surely O* transdures.

The problem is that END (6) merely tells us that the identity of an enduring object is grounded in the sameness, over time, of either a substratum or a qualitative property instance. But that characterisation is in fact consistent with an object transduring. Thus END (6) does not meet DISTINGUISHABILITY.

In the previous two sections we have seen that the metaphysical resources that Armstrong develops on behalf of the endurantist cannot do the work he hoped; for none of the resources it affords us allows us to distinguish endurance from transdurance. To do so we need to appeal to additional metaphysics. In what follows I follow Parsons (2007) and Eagle (2010) who develop a set of what they call location relations and then use these relations to better define the three competing views about persistence. In section 5 I argue that with these location relations to hand the endurantist need not appeal to any further metaphysical machinery to explicate her position: in particular she need not be committed to non-qualitative thisnesses or to substrata or to tropes and thus need not be committed to any ancillary metaphysical thesis to which the perdurantist is not committed. I return, in section 6, to the question of whether there are other reasons that the endurantist ought to be so committed.

5 Location relations

Location relations are relations that objects bear to regions of space-time. The assumption is that there exist regions of space-time and that objects are located at, or occupy, those regions. That is not a wholly uncontroversial assumption, so I do not suggest that explicating notions of persistence in this way incurs no metaphysical costs. The suggestion is merely that explicating the views in this manner allows us to make the distinctions we want and that the metaphysical cost – substantivalism about space-time – is not a high one. In what follows I borrow terminology from Eagle (2010). Eagle takes occupation to be a primitive. An object occupies any region where it can be found. That is, it occupies any region not free of it. Thus my dog, Annie, occupies the living room when she is sitting on the sofa in that room since one perfectly good answer to the question, where is Annie?, would be, the living room. We can then define three further notions.

> *Containment*: O is contained in R iff each part of O occupies a sub-region of R.
>
> *Filling*: O fills R iff each sub-region of R is occupied by O
>
> *Whole Location*: A whole location of O is any region, R, that both contains O and is filled by O, as long as no proper sub-region of R contains and is filled by O.
>
> *Exact Location*: O is exactly located at R iff O is contained in R and O fills R and no part of O occupies any region not overlapping R.

Put more informally an object, such as Annie, is *contained* in a region, R, just in case all of her parts occupy some sub-region of R. Thus when Annie is sitting on our sofa in the living room she is contained in the living room. When Annie is sitting on the balcony with one paw in the living room she is not contained in the living room. An object, such as Annie, fills a region just in case there is no sub-region that is not occupied by that object. Thus Annie *fills* the region her paw occupies. She also fills the dog-shaped region that she occupies. She does not fill the living room, since there are regions of the room at which she is absent. Now consider the location that an object, such as Annie, both fills and is contained in. If we think of Annie at some moment then this region is exactly dog-shaped. This is Annie's whole location since, in this case, there is no proper sub-region of that region that is also contained and filled by Annie. Annie's *exact location*, however, is not a three-dimensional one, at least on the assumption that she persists through time. Rather, it is the location that she both fills and is contained in – and it is thus a dog-shaped region (at least at a time) such that none of her parts occupies a region that does not occupy the region she exactly occupies. Thus if Annie lives from 2011 until 2031 then her exact location is a four-dimensional region that spans that time period, which has a proper sub-region each and every dog-shaped region she occupies at each moment during that period.

The notion of whole location is supposed to capture something like the notion of being wholly present at a time (or region). It is consistent with whole location, so defined, that something can be wholly located at different regions. Suppose Annie is an enduring object. Then she occupies a series of three-dimensional regions (Annie at each moment at which she exists) and she occupies a four-dimensional region (the fusion of those three-dimensional regions). In essence, it is the endurantist's contention that the way that Annie occupies the four-dimensional region is very different to the way she occupies each of the three-dimensional regions. For clarity, let us introduce a piece of terminology that allows us to pick out these three-dimensional regions. Following Eagle with can call these M-regions.

> M-region: An M-region is a maximally temporally unextended subregion of an exact location of an object O.

Each M-region corresponds to what we think of as the region occupied by an object at a moment of time. We can now spell out the two ways in which Annie occupies M-regions, on the one hand, and the four-dimensional region that is the fusion of those M-regions, on the other hand. If Annie endures then she is exactly located at the four-dimensional region. She is not exactly located at any M-region. Instead, Annie is wholly located at each M-region. But she is not wholly located at the four-dimensional region. Hence we can make sense of the idea that Annie is multiply wholly located: for she is wholly located at each M-region, but is exactly located at just one four-dimensional region.

On the other hand, suppose that Annie perdures. The perdurantist will agree with the endurantist that Annie is exactly located at some four-dimensional region. But she will disagree that Annie is wholly located at any M-region. Rather, the perdurantist thinks that Annie's whole location is the very same location as her exact location: the four-dimensional region. The perdurantist agrees with the endurantist that something is wholly located at each M-region: but she disagrees that that thing is Annie; instead she thinks that what is wholly located at each M-region is an instantaneous temporal part of Annie.

Finally, suppose that Annie transdures. According to the transdurantist both endurantists and perdurantists are right when they say that Annie is exactly located at the four-dimensional region. Further, the transdurantist thinks the perdurantist is right when she says that Annie is wholly located at the self-same region at which she is exactly located: the four-dimensional region. But the transdurantist disagrees with both the perdurantist and the endurantist, both of whom think that there is something wholly located at each M-region. For the transdurantist thinks there is nothing wholly located at any M-region occupied by Annie: neither Annie nor a temporal part of Annie. A transduring object occupies each M-region by filling that region, but transduring objects fail to be contained in such regions and thus fail to be wholly located at them. With these distinctions in mind we can define persistence as follows:

PERS: An object, O, persists iff it is exactly located at a four-dimensional region.

We can then define perdurance, endurance and transdurance in terms of the different relations that persisting objects bear to the regions they occupy.

PERD: O perdures iff (a) O is exactly located at a four-dimensional region, R, and (b) O is wholly located at R and (c) for any M-region of O there exists an O* such that (i) O* is wholly located at R* and (ii) O* is a proper part of O.

PERD entails that O persists since it is exactly located at a four-dimensional region. Moreover, it entails that O is not multiply located throughout that four-dimensional region since O is wholly located at that four-dimensional region (and thus cannot be wholly located at any sub-region of that region). Moreover, it tells us that for any M-region of O, something is wholly located at that region, and that thing is a proper part of O.

END: An object O, endures iff (a) O is exactly located at a four-dimensional region, R, and (b) for any M-region, R*, of O, O is wholly located at R*.

END entails that O persists since it is exactly located at a four-dimensional region. It is not, however, wholly located at that region since it is wholly located at each M-region. Thus enduring objects are multiply wholly located.

TRANS: An object O transdures iff (a) O is exactly located a four-dimensional region, R, and (b) O is wholly located at four-dimensional region, R, and (c) either (i) there is no object, O* that is wholly located at any M-region of O or (ii) any object, O* that is wholly located at any M-region of O region is not a part of O.[13]

TRANS is a little more complex. It entails that O persists, since it is exactly located at a four-dimensional region. It also tells us that O is wholly located at that same four-dimensional region. In this, the transdurantist agrees with the perdurantist. Moreover, the transdurantist agrees with the perdurantist that O is not wholly located at any sub-region of that four-dimensional region. Where she disagrees with the perdurantist and the endurantist lies in clause (c). The transdurantist might say that there is no object that is wholly located at any M-region of O. If so, she disagrees with both the endurantist and the perdurantist: for the endurantist thinks there is such an object – namely O – while the perdurantist thinks there is such an object – namely some instantaneous temporal part of O. This, I take it, is the most usual version of transdurantism. Alternatively the transdurantist might concede that there is such an object, O*, but maintain that O* is not O (since O is not wholly located at any sub-region of that four-dimensional region) nor any part of O (and thus not a temporal part of O). This latter option would have it that there are two objects, O and O*, each of which occupies the same M-region, but such that they are entirely distinct (they do not even overlap).[14]

What is clear is that we are able to define these three theories of persistence in a straightforward manner by appealing to location relations. Moreover, we need appeal to nothing more than those relations: there are no additional metaphysical claims to which the endurantist need be committed in order to explicate her view. Thus Armstrong's contention that occamist reasoning favours perdurantist is not, at least on these grounds, correct. We can make good sense of endurantism just by understanding the different ways that objects can (or could) (or hyperintensionally could) occupy regions.

---

[13] This is not the definition of transdurance that Daniels 2013, 2014 offers. He defines the notion in terms of Parsons' location relations and the definition is therefore not equivalent to this. Nevertheless, the view being defined is sufficiently similar to this view to warrant the same name.
[14] Something like this view is defended in Miller 2006.

6 Explanation and identity

Could one, however, make a case that Armstrong is right in his contention that occamist reasoning gives us reason to prefer perdurantism to endurantism. I am not sure, but in what follows I outline a line of thought that some might find compelling.[15]

Suppose one is an endurantist. Now ask the question that Armstrong asks: in virtue of what is the thing, O, that is wholly located in this region, the very same thing as the thing, O*, that is wholly located in that other region? Further, suppose one embraces Leibniz's Law. One might say that identity is a brute matter and in virtue of O and O* being identical, they share all the same properties. Or one might say that the identity of O and O* is grounded in the fact that they share all and only the same (temporally relativised) properties. In neither case need the endurantist posit additional ontology such as haecceities, thisnesses or special tropes that can be multi located and ground identity.

Many are happy with the idea of brute identity. I find it puzzling. So I would be tempted to ground the identity of O and O* in the fact that they share all of the same properties. But is that at all convincing? Consider a cross-world case rather than a cross-temporal case. Consider O in @ and O* in w* and O** in w*. O is red, then blue, then green. O* is black then purple then orange. O** is black, then purple, then orange. In fact, O* and O** are intrinsic duplicates of one another. Now suppose that we think that O is trans-world identical to objects in other worlds, (after all, we are supposing that we are endurantists) and stipulate that O is identical to one of O* or O** in w*. The brute identity theorist must say that it is simply a brute matter than O is identical to, say, O* rather than O**. The alternative view has it that O is identical to O* because O and O* share all the same properties. O has certain world-relativised properties: it has the property of being black-at-t-in-w* and being purple-at-t2-in-w*, and O* has the property of being red-at-t-in-@ and blue-at-t2-in-@. O** lacks the properties of being red-at-t-in-@ and blue-at-t2-in-@. Thus we could say that the identity of O with O* (and not O**) is grounded in them sharing the same temporally and modally relativised properties: they share a giant set of properties all of which are relativised to both times and worlds.

But of course, just as temporally indexed properties at-tn for some tn are invisible at all times other than tn, so too modally indexed properties at-@ are invisible at all worlds other than @. So one might think that such properties are odd things

---

15 The argument was made to me by David Braddon-Mitchell in conversation. I am not sure I find it wholly compelling, but for those that do it offers a reason to embrace Armstrong's Occamist conclusion.

to ground the identity of objects across worlds, or, indeed, across times. For there is nothing at all *intrinsically* about O* in virtue of which it is identical to O, and O** is not.

One might think that insofar as one is attracted to trans world identity theory one ought to embrace the existence of haecceities. Then one could say that what grounds the fact that O* has certain modally relativised properties – namely the properties of being certain ways in @ – is the fact that there is some haecceity in @ that has those properties and it is the *same* haecceity as in w*. That is, in answer to the question: but why does O* have the property of being blue-at-t1-in-@ and O** lacks that property, given that both of them have the very same set of intrinsic properties, we can point to some further feature of matters: a haecceity or thisness that O* and O share, and O** does not.

If one finds that line of thought compelling then one ought to find it equally compelling in the cross-temporal case. Consider O-at-t, the thing that is wholly located at some M-region R. Now consider O-at-t1, the thing that is wholly located at some M-region R1. There is nothing about the manifest, intrinsic, properties of O-at-t1 that suggests that it has the property of being a certain way at t, namely the way that O-at-t is. If we want to know why O-at-t1 has this set of temporally relativised properties, including all the properties that are relativised to times other than t1, then nothing about the intrinsic manifested features of O at t will help us. To say that O at t and O at t1 are one and the same thing that is wholly located at each region because they share the same set of relativised properties seems to be explanatorily unhelpful. But one might suggest that the reason that that cluster of properties goes together, as it were, is because there is some cross-temporal haecceity: a thisness that exists at different times and attracts the set of properties. What explains why those properties are instantiated at each time by each of the things is that each has the very same haecceity.

If one finds such an argument from explanation compelling, then one will think the endurantist ought to appeal to haecceities; and if she ought to appeal to haecceities then there are occamist grounds to prefer either perdurantism or transdurantism to endurantism. But the "ought'" at play here is a sort of explanatory ought: there is nothing about endurantism itself that demands one appeal to haecceities. Moreover, one might simply not feel the pull of the call for explanation here, and thus might be entirely happy to be a haecceity-less endurantist. I'm tempted to think that Armstrong would have been perfectly happy with that.

Bibliography

Armstrong, D. M. 1978. *Universals and Scientific Realism, Vol. II. A Theory of Universals*. Cambridge: Cambridge University Press.
Armstrong, D. M. 1980. "Identity Through Time". In van Inwagen, P. (ed), *Time and Cause: Essays Presented to Richard Taylor*. Dordrecht: Reidel, 67–78.
Audi, P. (forthcoming). "A Clarification and Defense of the Notion of Grounding" in Correia F., and Schnieder, B. (eds.), *Grounding and Explanation*, Cambridge: Cambridge University Press.
Balashov, Y. 1999. "Relativistic Objects". *Nous*, 33(4), 644–662.
Balashov, Y. 2000a. "Persistence and Space-time: Philosophical Lessons of the Pole and Barn". *The Monist*, 83(3), 321–240.
Balashov, Y. 2000b. "Relativity and Persistence". *Philosophy of Science*, 67(3), 549–562.
Balashov, Y. 2000c. "Enduring and Perduring Objects in Minkowski Space-Time". *Philosophical Studies*, 99, 129–166.
Balashov, Y. 2002. "On Stages, Worms and Relativity". *Philosophy*, 50, 223–252.
Balashov, Y. 2003. "Temporal Parts and Superluminal Motion". *Philosophical Papers*, 32(1), 1–13.
Benovsky, J. 2006. *Persistence Through Time and Across Possible Worlds*. Frankfurt: Ontos Verlag.
Benovsky, J. 2008. "The Bundle Theory and the Substratum Theory: Deadly Enemies or Twin Brothers?" *Philosophical Studies*, 141(2), 175–190.
Benovsky, J. 2010. "Relational and Substantival Ontologies and the Nature and the Role of Primitives in Ontological Theories". *Erkenntnis*, 73(1),101–121.
Braddon-Mitchell, D., and Miller, K. (2006). "Talking about a Universalist World". *Philosophical Studies*, 130(3), 507–542.
Braddon-Mitchell, D., and Miller, K. 2006b. "The physics of extended simples". *Analysis*, 66(3), 222–226.
Braddon-Mitchell, D., and Miller, K. 2007. "There is no simpliciter simpliciter". *Philosophical Studies*, 136(2), 249–278.
Chisholm, R. M. 1973. "Parts as Essential to Their Wholes". *Review of Metaphysics*, 26, 581–603.
Chisholm, R. M. 1975. "Mereological Essentialism: Further Considerations". *Review of Metaphysics*, 28, 477–484.
Chisholm, R. M. 1976. *Person and Object. A Metaphysical Study*. La Salle (IL): Open Court.
Daniels, P. 2013. "Occupy Wall: A Mereological Puzzle and the Burdens of Endurantism". *Australasian Journal of Philosophy*, 92(1),1–11.
Daniels, P. 2014. "The Persistent Time Traveller: Contemporary Issues in the Metaphysics of Time and Persistence". PhD Thesis.
Eagle, A. 2010. "Location and Perdurance". In Zimmerman, D. (ed.), *Oxford Studies in Metaphysics. Vol. 5*. New York: Oxford University Press, 53–94.
Haslanger, S. 1989. "Endurance and Temporary Intrinsics". *Analysis* 49, 119–125.
Hawley, K. 1998. "Why Temporary Properties Are Not Relations Between Physical Objects and Times". *Proceedings of the Aristotelian Society*, 98 (2), 211–216.
Hawley, K. 2001. *How Things Persist*. Oxford: Oxford University Press.

Heller, M. 1984. "Temporal Parts of Four-Dimensional Objects". *Philosophical Studies*, 46, 323–334.
Lewis, D. K. 1983. "Survival and Identity". In Id. 1983, *Philosophical Papers, Vol. 1*. Oxford: Oxford University Press, 55–77.
Lowe, E. J. 1987. "Lewis on Perdurance Versus Endurance." *Analysis*, 47: 152–54.
Merricks, T. 1994. "Endurance and Indiscernibility". *The Journal of Philosophy*, 91, 165–84.
Merricks, T. 1999. "Persistence, Parts and Presentism". *Nôus*, 33(3), 421–438.
Miller, K. 2005. "The Metaphysical Equivalence of Three and Four Dimensionalism". *Erkenntnis*, 62(1), 91–117.
Miller, K. 2006. "Non-mereological Universalism". *European Journal of Philosophy*, 14(3), 427–445.
Miller, K. 2009. "Ought a four-dimensionalist to believe in temporal parts?" *Canadian Journal of Philosophy*, 39(4), 619–647.
Parsons, J. 2004. "Distributional Properties". In Jackson F., and Priest, G. (eds). *Lewisian Themes: The Philosophy of David K. Lewis*. Oxford: Clarendon Press.
Parsons, J. 2007. "Theories of Location". In Zimmerman, D. (ed.), *Oxford Studies in Metaphysics. Vol. 3*. New York: Oxford University Press, 201–232.
Schaffer, J. (2009). "On What Grounds What". In Chalmers, D., Manley, D., and Wasserman R. (eds.), *Metametaphysics*. Oxford: Oxford University Press,347–383.
Sider, T. 2001. *Four-Dimensionalism: An Ontology of Persistence and Time*. Oxford: Clarendon Press.
Simons, P. 2004. "Extended simples: a third way between atoms and gunk". *The Monist*, 87(3): 371–385.
Thomson, J. J. 1983. "Parthood and Identity Across Time". *Journal of Philosophy*, 80, 201–220.
Thomson, J. J. 1998. "The Statue and the Clay". *Nôus*, 32, 149–173.
Trogdon, K. (forthcoming). "An Introduction to Grounding". In Hoeltje, M., Schnieder B., and Steinberg A. (eds.), *Basic Philosophical Concepts*, Munich: Philosophia Verlag.
Van Inwagen, P. 1987. "When are Objects Parts?". *Philosophical Perspectives, 1. Metaphysics*, 21–47.
Van Inwagen, P. 1990. "Four-Dimensional Objects". *Nôus*, 24, 245–255.

Stephen Mumford
Armstrong on Dispositions and Laws of Nature

1 Dispositions, ontologically speaking

Armstrong did not take dispositions to be as real and full-bodied as some of us would like. Nevertheless, he made real progress in understanding them ontologically. Before his account appeared (Armstrong, 1968, 85–8), Ryle's phenomenalism about dispositions was still orthodox. Ryle's position was so named because of its similarity with the better-known phenomenalism about material objects. The latter view, associated with the work of Berkeley, is one that Armstrong knew well from his early work (see Armstrong 1960 and Martin and Armstrong, eds, 1968) so he was perfectly placed to see the parallels. To say that some particular thing is soluble, according to Ryle's view (Ryle 1949, ch. 5), is just to say that if it were to be in liquid, then it would dissolve. This Rylean conditional analysis is squarely in the empiricist tradition. It refuses to take dispositions seriously – as real existents – and instead interprets their ascriptions to be reducible to conditional claims about observable occurrences. While Ryle's philosophy was radical in many respects, his treatment of dispositions would not have been out of place in Hume's *Treatise* (Hume 1739).

Armstrong was impressed by Martin's truthmaker principle: a principle first developed in relation to Berkeley's phenomenalism (Armstrong 1989, 8–11). According to Berkeley, the claim that there is a chair in your living room means nothing more than that if you were to enter your living room, you would have chair-like sensations. Material objects thereby get reduced to sensations and, when unperceived, to the possibility of sensations. We could rightly ask, however, on what basis we should accept the truth of conditional statements of the kind Berkeley proposed. What makes it true that if you enter your room, you will have a sensation of the chair-kind? There is an obvious answer. The chair – a material object – exists, is placed in your room, and makes it true, even when unperceived, that if you were to enter the room you could have chair-like sensations. But this answer is exactly what Berkeley was seeking to avoid. His project was to reduce the material to the mental and the project fails if we have to appeal to mind-independent objects as the truthmakers for the analysis.

Stephen Mumford: University of Nottingham and Norwegian University of Life Sciences (NMBU), email: stephen.mumford@nottingham.ac.uk

Armstrong saw that exactly the same type of argument could be brought to bear against Ryle's theory of dispositions. What makes it true that if this soluble sugar cube is placed in liquid, it would dissolve? An obvious answer suggests itself: the sugar cube is soluble. It has a property that is causally responsible for it dissolving under certain conditions. And yet, again, this is the very sort of answer that Ryle sought to avoid by offering his conditional analysis of disposition ascription. Ryle wanted nothing more than the conditionals, whose antecedent and consequent conditions were perfectly occurrent terms. In that way, there was no need to allow dispositions as real properties or states. But one merely has to ask the truthmaker question – what is the truthmaker of such a conditional – to see that a simple conditional analysis leaves everything unexplained and therefore offers no account at all of why we should accept the truth of such conditionals.

Instead, Armstrong offered what he called a 'realist' account of dispositions (Armstrong 1968, 86). There was, he argued, a categorical basis for each disposition, which was causally responsible for the production of certain manifestations under certain conditions. The basis persisted between those manifestations of the disposition and was thereby able to act as truthmaker for any counterfactual conditionals that is true while the disposition is unmanifested. Indeed, such categorical bases could be gained, and lost, and gained again, without any stimulus condition being applied to the disposition. In that case, the associated counterfactual condition could become true, then false, then true again, even though there is never a test applied for the presence of the disposition. Armstrong's account is realist in this sense and, while I will argue that it is not nearly realist enough, we ought to acknowledge it as one of the first contemporary accounts that tried to understand dispositions ontologically instead of, like Ryle, seeking an empiricist and reductive account of their being. Armstrong's 'realist' theory precedes those of Bhaskar (1975) and Harré and Madden (1975), for instance, and is influenced more by metaphysical considerations than the philosophy of science which was behind those later accounts.

2 Was Armstrong's account sufficiently realist?

Armstrong's metaphysics can be seen as a serious effort to restore realist metaphysics after a lengthy reign of reductive empiricism. Hume's ideas persisted into the Twentieth Century, manifest in an extreme form in logical positivism (see Hanfling, ed. 1981), and exerting a continued influence through the neo-Humean programme of David Lewis (1973, 1986). Armstrong's work is anti-Humean in a number of respects but, I will argue, not enough. Armstrong failed to see through the realist programme in the fullest sense and his final stance was a half-way position. It was not quite neo-Humean but not quite neo-Aristotelian either. Instead, Armstrongian metaphysics has an element of both. Whether it is possible to sustain such a compromise position is a matter that I cannot issue definitive judgement upon here. My aim, rather, is to explain why I understand Armstrong to have this midway position and point to some of the tensions that are created by attempting to hold a middle ground. One serious question, for instance, is whether there is any middle ground to occupy at all or, rather, whether his final position is simply Humean on some points and anti-Humean on others. Given that Humean and Aristotelian metaphysics can both be considered as coherent wholes, then occupancy of an intermediate position looks unstable from the start. It will be clear that I am sceptical of this attempted compromise but I accept that someone might in the future be able to produce a satisfactory account of the kind Armstrong sought.

I will now add some substance to these broad-brush statements. The starting point in this matter is that anti-Humeans are realists about powers. Although Armstrong initially called his a realist theory of dispositions, as the debate played out it became clearer, including to Armstrong, that he was not on the side of powers in this respect. His realism is not an acceptance of an ontology of irreducible powers in the full sense which many pre-Moderns and contemporary Aristotelians accept. And this is betrayed by the fact that Armstrong still needed laws of nature in his metaphysics, filling the explanatory gap that is left once real powers and their associated final causes are removed. Of course, the powers of scholastic philosophy (as detailed by Feser 2014, for instance) have long been objects of suspicion in Modern philosophy, when all of Aristotle's four causes except efficient causation (see Schmaltz ed. 2014) were jettisoned. The idea that real, irreducible powers or natures tend towards their manifestations was dropped and in its place we saw the rise of laws of nature governing the behaviour of things. We can have a powers-based metaphysics or a laws-plus-categorical properties one, and Armstrong clearly commits to the latter (see his 2005, 315; 1989, 15; 1996, 17 and Mumford 2007, 92). As Armstrong says, laws are the truthmakers for the attributions of powers. For him, laws are the existents. Disposition-talk can be useful but the metaphysics behind it is one of categorical properties and laws of nature.

More recent philosophy of science and metaphysics has seen a revival of interest in the metaphysics of active powers in which things contain an internal impetus to change in virtue of their natures and do not then need to be governed from without. As well as Bhaskar and Harré and Madden, we can think of Ellis (2001), Mumford (2004), Mumford and Anjum (2011), Groff (2013) and Marmodoro (2015) as developing such realist views, which go way beyond Armstrong's own realism about dispositions. Nevertheless, we have to acknowledge that Armstrong was very much aware of the powers view and his rejection of it was principled and reasoned. We should therefore consider why he continued to resist it.

3 Powers, actualism and degrees of being

Armstrong does not accept the reality of bare powers. He does not accept, then, pandispositionalism: the view that all properties are powers. But he does not even accept that some properties are irreducibly dispositional as all must be ultimately grounded in their categorical bases.

A major reason Armstrong gave consistently against acceptance of pandispositionalism is an argument he attributed to Swinburne (1983). This argument was named by Molnar (2003, 173) the Always Packing, Never Travelling argument. Armstrong articulated it thus:

> Can it be that everything is potency, and act is the mere shifting around of potencies? I would hesitate to say that this involves an actual contradiction. But it does seem to be a very counter-intuitive view. The late Professor A. Boyce Gibson, of Melbourne University, wittily said that the linguistic philosophers were always packing their bags for a journey they never took. Given a purely Dispositionalist account of properties, particulars would seem to be always re-packing their bags as they change their properties, yet never taking a journey from potency to act. For 'act', on this view, is no more than a different potency. (Armstrong 1997: 80)

In later statements, he offered some clarification on the meaning of this charge: 'Causality becomes the mere passing around of powers from particulars to further particulars ... the world never passes from potency to act.' (Armstrong 2005, 314, and see also Armstrong 1983, 123)

We may reconstruct the argument as follows. Suppose all properties were powers. A power is to be understood as being for a particular actualisation, sometimes known as its manifestation. But, given that under the assumption of pandispositionalism all properties are powers, then the manifestation of a power can only be in another power. So the apparent actualisation of a power is merely in the possession of a new power, that is, something which also awaits an actualisation. And Armstrong clearly thinks we must conclude that nothing thereby gets actualised. Of course it might be said that something nevertheless does happen. If a power has been passed on, then that sounds like an occurrence: a passing on. But given that what is passed on is something that is non-actualised, then Armstrong clearly sees this as not a real happening in its own right at all. *Something* being passed around would count as a real happening. But a non-actualised power does not count as *something*, for Armstrong, and thus the passing around of such non-things does not count as a real happening either.

There have been various attempts to answer the Always Packing, Never Travelling objection (for example, Molnar 2003, ch. 11 and Mumford 2004, 174–5). A different take on the issue has been suggested recently by Feser (2014, 85–6), which I adapt and add my own comments towards here. There seems to be an underlying basic assumption in Armstrong's position, which, once made, leads us inevitably towards the conclusion that bare powers have no existence. This is the assumption of Andersonian naturalism, almost certainly taken by Armstrong from the influential Australian philosopher John Anderson. The key facet of Andersonian naturalism, for present purposes, is that there are no degrees of being. Everything either exists, and is actual, or does not and is not. J. L. Mackie summed up the Andersonian position simply:

> His central doctrine is that there is only one way of being, that of ordinary things in space and time, and that every question is a simple issue of truth or falsity, that there are no different degrees or kinds of truth. (Mackie 1962: 265)

Armstrong has made a number of pronouncements along these lines, committing to a very similar kind of naturalism, for example, speaking directly of Anderson:

> Anderson held that the world was the spatio-temporal world, and that nothing else existed except this world. Not only was there no God, or non-spatial minds, but there were no 'abstract' entities in the Quinean/North American sense of that term: entities over and above the spatio-temporal world. So among the other things which Anderson excluded, there were no Platonic forms or realm of universals descried by the eye of reason. Realism about universals for Anderson meant that different things in the spatio-temporal world could have the *same* quality or property, or be of the *same* kind or sort. It was a thoroughly down to earth (down to space-time) form of realism. (1984: 41–2)

Armstrong could distil his naturalism into a simple statement: 'Naturalism I define as the doctrine that reality consists of nothing but a single all-embracing spatio-temporal system.' (Armstrong 1981, 149)

Now the problem this view creates for powers is that it suggests a rather sparse ontology. Specifically, Armstrong accepts a simple division between what is actual and what is not. The actual is equated with the sum of existence and the non-actual is equated with the non-existent. This view can be called actualism, and Armstrong has a clear commitment to it, for example:

> I assume the truth of what may be called Actualism. According to this view, we should not postulate any particulars except actual particulars, nor any properties and relations (universals) save actual, or categorical, properties and relations. I do not think this should debar us from thinking that both the past and the future exist, or are real. But it does debar us

from admitting into our ontology the merely possible, not only the merely logically possible but also the merely physically possible. ... This debars us from postulating such properties as dispositions and powers where these are conceived of as properties over and above the categorical properties of objects (Armstrong 1983, 8–9).

We might call this the Armstrongian cleavage between what is and what is not. But it is clear that those who are realists about powers have a richer conception of reality: one which applies a different cleavage and in which there can be some things that are existent without being actualised or in act. Powers exist in potency, to use the Aristotelian terminology (see Feser 2014, ch. 1). Hence, the neo-Aristotelian effectively has a threefold distinction: between what is non-existent, what is existent as potency, and what is existent as act. How this maps on to Armstrong's division is show in figure 1.

Armstrongianism	Aristotelianism
Existent/actual	Actuality
	Potency
Non-existent/non-actual	Non-existent

Fig. 1: *The Armstrongian and Aristotelian cleavages*

The effect of Armstrong's twofold distinction is that an unmanifested power is taken to be a mere possibility of actuality, where mere possibilities for Armstrong have no existence at all. The passing around of powers is thus not a real passing, for nothing is passed. The neo-Aristotelian has a different starting point altogether. An unmanifested power is real, *qua* potency, even if it is not actualised in its manifestation. But such potencies are very much existent even when not manifested. Hence, the passing around of power is the passing of something real. For an Aristotelian, then, the world of powers is not a world lacking in existence. This is not to say that there are no more questions for pandispositionalism to answer but at least one objection to the view is disarmed by this different starting point with its richer conception.

4 Potency and act

Another way of understanding the difference between these two ontologies is in their treatments of possibility. Armstrong set aside the possibilities that are also actualities – for, trivially, that which is actual is also possible – and he called the rest *mere* possibilities. For Armstrong, mere possibilities have no existence at all. You shouldn't fear a possible murderer, for instance; only a real one can kill you. Armstrong emphasised at various times fictionalist and combinatorialist accounts of these mere possibilia (Armstrong 1989). They were no part of reality although they could be reconstructed, mentally for instance, as fictional recombinations of the elements within reality. The Aristotelian makes a distinction, however. A potency has real being. It is a thing's nature: for instance to dispose towards a type of manifestation. It is in the nature of sugar to dissolve in liquid, for example. Now there are many things which are merely possible but for which there is no real potency. Hence, it is a mere, non-existent possibility that there be a turquoise hippopotamus; but hippos have no natural disposition towards this colour. Such a possibility would thus have no real worldly existence. A particular female hippo's potential to be pregnant is, in contrast, a part of reality, even if she never becomes pregnant. She has a real potency towards child-bearing. Armstrong's equation of merely possible with non-existent does not hold straightforwardly for realists about powers, therefore. Some, though not all, of what is potentially actual is grounded in real potency.

It might nevertheless be thought the same objection, of Always Packing, Never Travelling, can be reasserted in a modified form. Perhaps one can admit that unmanifested powers are real, qua real potencies, but there is nevertheless an outstanding question concerning their actualisation. One might then state the Swinburne/Armstrong argument in terms, as Armstrong does, of never making the journey from potency to act. Armstrong seemed to think that in failing to do so, powers never passed into reality, and we have already seen how this conclusion can be resisted. But there is still the problem for the dispositionalist that the existence of powers might always be existence qua potency and they can never achieve existence qua actuality. The reason for this is that the actualisation of each power would only be in another power: something that exists only in potency until it is actualised. Given that the same argument applies to the subsequent power, then it seems that every power is in want of an actualisation.

The pandispositionalist does, however, have a way of explaining away this apparent problem. When a power manifests itself, it is indeed actualised even though, for a pandispositionalist, that actualisation will also, at the same time, create a new unmanifested power. The following example should help explain this. Suppose something had a power to become spherical or to create a spherical thing. This might be the power of a certain plant to grow pea pods, in which are contained spherical peas. The sphericity of the peas of the pod, to begin with, is potential only, grounded in the real potency of the plant. This potency will attain its actualisation only when a spherical pea or peas are produced. Assume now that this indeed comes to pass. When it does so, this particular potency has been actualised. But, in the pandispositionalist view, every property is a power. Sphericity, for example, can be understood as a power to roll in a straight line on an inclined plane. So the actualisation of one power – its end state – is in the creation of a new power existing in potency. One power has done its work in producing a change, and has thus been actualised, even though its actualisation was the creation of a new power in potency. Such an account has been produced by Mumford and Anjum (2011). Marmodoro has forced the issue of whether the manifestation of a power is the creation of a new power or a different state of the same power (Marmodoro 2013: 550). She thinks the latter is the case though she says that we (Mumford and Anjum) prefer the former. But the above account shows how the manifestation of a power can be both the creation of a new power and the end state of the first power. The end of one power – its final manifestation – is precisely the creation of a new power, for a pandispositionalist.

5 Laws to the rescue?

We have seen how the Aristotelian view, which persisted into the medieval period in the philosophy of Aquinas (see *Summa Contra Gentiles* III, ii for the best Thomistic statement) had a rich ontology of powers. Modern philosophy spurned this, and the notion of cause that was retained was the one that corresponded to Aristotle's efficient causation alone. In particular, there were no final causes, so no sense of a power acting for an end. This latter notion is crucial in the powers metaphysic, which speaks of powers that naturally dispose or tend towards their final manifestation. Many contemporary realists about powers try to avoid appeal to final causes, perhaps because of their perceived ill-repute. But this usually results in them trying to say similar things in other words: using an intentional or normative vocabulary where a dispositional one would do better.

Armstrong's own solution, while he contrasts it with Humeanism, is clearly in agreement with the Modern approach. He rejects powers and instead adds laws to govern the workings of nature. It seems undeniable that there is a degree of order in the world. Heated iron bars expand, struck matches light, gelignite explodes, and so on. Armstrong has denied irreducible powers and instead rules that all properties are categorical or non-dispositional. He calls this position Categoricalism. Categorical properties do not intrinsically dispose towards certain ends. But Armstrong instead explains the degree or order in reality with his theory that properties, qua immanent universals, partake in laws of nature. Categorical properties are indeed the relata of nomological relations.

The theory is presented as anti-Humean for good reason. The Humean view takes laws to be simply the contingent regularities that there happen to be: either simple regularities or, for Mill (1843, III, iv, 1), Ramsey (1929, 150) and Lewis (1973, 73), the regularities that would be the axioms or theorems of the best possible systematisation of the world's history. There are at least two major problems with the regularity view, which Armstrong (1983, ch. 4) presents. Such Humean laws do not explain nor govern their instances because there are merely descriptive of, or even constituted by, those instances. And, second, such laws would not support counterfactual instances. The gravitation law, for example, does not just apply to all actual material bodies but to anything else were it to be a material body. Suppose there were one extra planet in our solar system, for example. The gravitation law would apply to it, we assume. Strictly speaking, a Humean has no simple principled reason for saying so.

Armstrong's solution, as is well known, is to take a law of nature to be a higher-order relation of natural necessitation holding directly between universals rather than between the particular things bearing those universals. The form of a law statement is thus N(F,G). And because all the instances of a universal are identical, then it follows that anything that is F will be G, meaning that N(F,G) entails the regularity $\forall x(Fx \rightarrow Gx)$. The law, in that sense, governs the instances and supports counterfactuals .If anything that is not F, were F, then it too would be G. The nomic relation is instantiated in particular things being F causing them also to be G: for example, in the heating of an iron bar causing it to expand. Hence, the law explains its instances instead of being explained by them. As can be seen, the theory is also offered as an explanation of dispositionality. While we may say that something, a, has a power, P, directed towards Q, Armstrong's theory tells us that a actually has a categorical property C and there is a law of nature N(C,Q) that explains the disposition without invoking primitive powers.

But this account, though anti-Humean to an extent, still should not be accepted by those who are serious about powers. One problem is that it retains the Humean assumption that laws are about exceptionless regularities and it thus fails to capture what is authentically dispositional about the dispositionalist's position: namely that natures tend, and no more than *tend*, towards their manifestations. Thus, Armstrong rejects Hume's account for saying that a law is a regularity, but in his own theory it is still the case that the existence of a law entails an associated regularity.

6 Tendencies

Against this, the Aristotelian tradition emphasises *what tends to be*. The reason is that powers can cut across each other, preventing or interfering with one another's full effects. Where we have counteracting powers, we can still have a real tendency towards G being exercised but, even if it is, then there is no guarantee that G will come about. J. S. Mill was one of the first philosophers in the empiricist tradition to rediscover this insight of ancient and medieval metaphysics. He explained the situation thus:

> To accommodate expression of the law to the real phenomena, we must say, not that the object moves, but that it *tends* to move, in the direction and with the velocity specified. We might, indeed, guard our expression in a different mode, by saying that the body moves in the manner unless prevented, or except in so far as prevented, by some counteracting cause. But the body does not only move in that manner unless counteracted; it *tends* to move in that manner even when counteracted; it still exerts, in the original direction, the same energy of movement as if its first impulse had been undisturbed, and produces, by that energy, an exactly equivalent quantity of effect. (Mill 1843, III.x.5, 444–5, italic in original)

He went on to conclude:

> These facts are correctly indicated by the expression tendency. All laws of causation, in consequence of their liability to be counteracted, require to be stated in words affirmative of tendencies only, and not of actual results. (Mill 1843, III.x.5, 445)

A number of contemporary philosophers of science accept this view or at least something in sympathy with it. However, even here the commitment to the metaphysical irreducibility of dispositionality is not explicit and, instead, there is an attempt to capture a *sui generis* modality of *disposing* without mention of it. Consider Cartwright, for example, who says the following:

> To ascribe a behaviour to the nature of a feature is to claim that that behaviour is exportable beyond the strict confines of the *ceteris paribus* conditions, although usually only as a 'tendency' or a 'trying'. ... The point here is that we must not confuse a wide-ranging nature with the universal applicability of the related law. To admit that forces tend to cause the prescribed acceleration (and indeed do so in felicitous conditions) is a long way from admitting that $F=ma$, read as a claim of regular association, is universally true. (Cartwright 1999, 28–9, see also 82)

Cartwright here uses first an intentional term, 'trying', and then a normative term, 'prescribed', putatively in explanation of the connection between the tendency and its manifestation. But the ideas of disposing towards or tending towards look perfectly adequate and the reluctance to speak in such ways seemingly betrays a reticence to appeal to a notion of final causes. Whatever the case may be in that respect, I am arguing that the idea of tending or disposing towards a manifestation is the crucial one that distinguishes dispositionalism from Humeanism (Mumford and Anjum 2011a). And Armstrong, anti-Humean to an extent, seems to side with Humeanism on this issue.

Now one can hardly deny that causal processes are capable of prevention and counteraction. You can stop a struck match from lighting, for instance, by blowing on it at the same time. To think of this in terms of laws, one seems forced to accept that laws admit exceptions. But this is far from easy for any conventional theory of laws. The regularity account is that a law is a constant conjunction so it would be a major amendment to allow that this conjunction can sometimes be less than constant. And for Armstrong's theory, exceptions seem to bring a particular difficulty. Universals are supposed to be identical in their instances, so one shouldn't really have one instance of F that causes something to be G and another instance of F that doesn't. The issue is pressing, therefore.

Armstrong has two possible responses. The first is to appeal to a distinction between iron and oaken laws (Armstrong 1983, 147–50). An oaken law is one such as N(F,G) but where there exists some further property, call it O, that, when added to F, F and O together necessitate another property H, where G is incompatible with H (incompatible, for instance, by being different same-level determinates under the same determinable). Hence, it can be the case that N(F,G) and N(F & O, H). (One would like to say that the law is N(F & O, ¬G), which would make the point clearer except that Armstrong (1978, ch. 14) does not allow the real existence of negative properties so there is no law that features a property such as ¬G). Iron laws are truly exceptionless in the sense that there are no counter-instances.

The notion of an oaken law does not solve the problem, however: at least, such an account is not able to provide what the tendency view wants. With an oaken law, one has to list all the additional properties that are possible preventers of, or interferers with, G, given F. There would have to exist a law for each possible interferer, leading to a multiplication of laws, arguably beyond necessity. The notion of tendency, in contrast, has an open-endedness that serves better. Why suppose, for example, that the possible interferers with respect to an F's production of G are finite in number? To say that being F tends towards being G captures the required force adequately and simply, whereas the attempt to preserve regularity as the basis of nature leads to us chasing ever more and more laws. It remains possible, as well, that N(F & O, H) could itself be interfered with, by adding another factor that again results in G. This is reflected in the non-monotonic nature of causal reasoning, which allows us to maintain all of:

1. Being F disposes towards G
2. Being F & O disposes towards ¬G
3. Being F & O & P disposes towards G
   ...and so on

In that case, why say that F & O necessitate H (or ¬G) when H could be prevented by the addition of P? Why not allow that F and O only tend towards H? This open-endedness is allowed by tendencies. Tending towards a manifestation is consistent with failing to reach it due to the interference of countervailing powers. Any specification of interfering conditions in a finite list, as in Armstrong's oaken laws, retains the possibility of falsehood through further interference.

Armstrong has a second strategy that is relevant. He allows that some laws are probabilistic. Might these be laws of tending towards an outcome instead of guaranteeing it? In that case, perhaps probabilistic laws explain why there are some regularities that are less than constant conjunctions. A probabilistic law $Pr:\frac{1}{2}(H,I)$, for example, could make it that half – instead of all – of those things which are H are also I. One might even wonder whether all powers are simply probabilistic propensities towards their effects.

But we should reject this interpretation of tendencies. For one thing, probabilistic laws are presumably bound by the axioms of the calculus of probability in which there can never be a > 1 chance of an occurrence. But tendencies, while coming in degrees of strength, are not so bound. There are ample mundane cases of overdisposing, for instance, where there is more than enough for an effect to occur (Mumford and Anjum, forthcoming). To kill a deadly bacterium, for instance, a medic might administer a dose that is more than enough to succeed. If the consequence of failing to eliminate the bacterium is dire, one might sensibly give twice an adequate dosage, just to take no chances. So tendencies are not mere probabilities.

We can add to this, however, that Armstrong's own particular account of probabilistic laws is inadequate to capture the sense of tendency required by the dispositionalist. Armstrong's laws work through necessity, as made clear when the simple necessity relation N relates the universals F and G. But the probabilistic case is not that much different, in Armstrong's theory. Even though Armstrong allows that some laws are probabilistic, on further analysis he says that these concern the probabilities of necessitation (Armstrong 1983, 132). So $Pr:\frac{1}{2}(H,I)$ means that there's a half chance that being H will necessitate being I. A probabilistic law is simply a special case where there's a < 1 chance of one universal necessitating another. But where an instance of H indeed produces an instance of I, it always does so through necessitating it.

In contrast, the tendency view rejects this basic principle. The idea that causal production works through necessitation of the effect by the cause is a metaphysical thesis designed to deliver constant conjunction. Once that requirement is removed, then we can divorce the concepts of causal production and causal necessitation. The dispositionalist accepts the former without the latter. Causes can be understood as tending towards, and no more that tending towards, their effects; sometimes producing them. A probabilistic law will not be able to give us the same as the notion of tendency, therefore. Indeed, the dispositionalist will want to apply his or her account even to probabilistic laws: these will tend, and no more than tend, towards a distribution of events. Armstrong's law Pr:½(H,I), if there be such a thing, will only tend towards a distribution of half the things that are H also being I. There is nothing that necessitates such a ratio.

There is a final reason why the defender of powers will not accept Armstrong's conciliatory brand of anti-Humeanism. Armstrong makes a big concession to his opponent when he accepts that it is, in a sense, contingent what causes what. Although each law works through necessity, as it involves a relation of natural necessitation, it is contingent what is lawfully related to what. Thus, while it may be a law that N(F,G), it could instead – in the combinatorial sense of could – be the case that N(F,H), where G is incompatible with H. This means that F could have a different nomological and causal role to the one it actually has and yet still remain F. A dispositionalist sees the nature of a property to be exhausted by its dispositional role. In no sense could it be allowed that a property F change that role and still remain the property it is. The power to roll in a straight line is thus an essential part of what it is to be spherical. If this disposition is lost, or swapped with another property, then that property is no longer sphericity.

7 Conclusion

Despite Armstrong's initial characterization of his position as a realist theory of dispositions, those who are serious about a metaphysics of powers have ample grounds to reject Armstrong's approach. The categoricalist view he offered, supplemented with a theory of laws of nature, does not provide what a thoroughgoing dispositionalist wants. I have tried to attend to the key matters of disagreement. Armstrong does not admit the reality of unactualised powers because he has a more fundamental actualist commitment. To that extent, he is anti-Aristotelian. The dispositionalist also has reasons to resist the law-governed account of nature. Such an account concedes too much to a Humean conception of reality: one understood in terms of exceptionless regularities instead of tendencies. Armstrong, and many others, are reluctant to accept the latter, especially if it involves a notion of final cause. However, it looks as if the attempt to fill the explanatory gap that is left, if one rejects finality, cannot do the job as well. To that extent, it looks very difficult to hold the sort of half-way anti-Humeanism metaphysic that Armstrong proposed.

Bibliography

Aquinas, St. T. *Summa Contra Gentiles*, trans. V. J. Bourke, New York: Doubleday, 1956.
Armstrong, D. M. 1960. *Berkeley's Theory of Vision*. Melbourne: Melbourne University Press.
Armstrong, D. M. 1968. *A Materialist Theory of The Mind*. Rev edn, London: Routledge, 1993.
Armstrong, D. M. 1978. *A Theory of Universals*. Cambridge: Cambridge University Press.
Armstrong, D. M. 1981. *The Nature of Mind and Other Essays*. Brighton: Harvester Press.
Armstrong, D. M. 1983. *What is a Law of Nature?* Cambridge: Cambridge University Press.
Armstrong, D. M. 1984. "Self Profile". In Bogdan, R. (ed.), *David Armstrong*. Dordrecht: Reidel, 3–51.
Armstrong, D. M. 1989. *A Combinatorial Theory of Possibility*. Cambridge: Cambridge University Press.
Armstrong, D. M. 1996. "Dispositions as Categorical States". In Crane, T. (ed.), *Disposition: A Debate*. London: Routledge, 15–18.
Armstrong, D. M. 1997. *A World of States of Affairs*. Cambridge: Cambridge University Press.
Armstrong, D. M. 2005. "Four Disputes About Properties". *Synthese*, 144, 309–20.
Bhaskar, R. 1975. *A Realist Theory of Science*. Leeds: Leeds Books Limited.
Cartwright, N. 1999. *The Dappled World*. Cambridge: Cambridge University Press.
Ellis, B. 2001. *Scientific Essentialism*. Cambridge: Cambridge University Press.
Feser, E. 2014. *Scholastic Metaphysics*. Heusenstamm: Editiones Scholasticae.
Groff, R. 2013. *Ontology Revisited*. London: Routledge.
Hanfling, O. (ed.) 1981. *Essential Readings in Logical Positivism*. Oxford, Blackwell.
Harré, R. and Madden, E. H. 1975. *Causal Powers: A Theory of Natural Necessity*. Oxford: Blackwell.
Hume, D. 1739. *A Treatise of Human Nature*, L. A. Selby-Bigge (ed.). Oxford: Clarendon Press, 1888.
Lewis, D. K. 1973. *Counterfactuals*. Oxford: Oxford University Press.
Lewis, D. K. 1986. *On The Plurality of Worlds*. Oxford: Blackwell.
Mackie, J. L. 1962. "The Philosophy of John Anderson". *Australasian Journal of Philosophy*, 40, 265–822.
Marmodoro, A. 2013. "Causes as Powers". *Metascience*, 22, 549–554.
Marmodoro, A. 2015. "Aristotelian Powers at Work: Reciprocity Without Symmetry in Causation". In Jacobs, J. (ed.), *Putting Powers to Work: Causal Powers in Contemporary Metaphysics*. Oxford: Oxford University Press, 54–73.
Martin, C. B. and Armstrong, D. M. (eds) 1968. *Locke and Berkeley: a Collection of Critical Essays*. Notre Dame: University of Notre Dame Press.
Mill, J. S. 1843. *A System of Logic, Collected Works of John Stuart Mill*, v. 7. Toronto: University of Toronto Press, 1973.

Molnar, G. 2003. *Powers: A Study in Metaphysics*. S. Mumford (ed.), Oxford: Oxford University Press.
Mumford, S. 2004. *Laws in Nature*. London: Routledge.
Mumford, S. 2007. *David Armstrong*. Chesham: Acumen.
Mumford, S. and Anjum, R. L. 2011. *Getting Causes from Powers*. Oxford: Oxford University Press.
Mumford, S. and Anjum, R. L. 2011a. "Dispositional Modality". In Gethmann, C. F. (ed.), *Lebenswelt und Wissenschaft: Deutsches Jahrbuch für Philosophie 2*. Hamburg: Meiner Verlag, 468–482.
Mumford, S. and Anjum, R. L. forthcoming. "Overdisposed". In *What Tends to Be: Essays on the Dispositional Modality*.
Ramsey, F. P. 1929. "General Propositions and Causality". In Mellor, D. H. (ed.), *F. P. Ramsey: Philosophical Papers*. Cambridge: Cambridge University Press 145–163.
Ryle, G. 1949. *The Concept of Mind*. London: Hutchinson.
Schmaltz, T. (ed) 2014. *Efficient Causation: A History*. Oxford: Oxford University Press.
Swinburne, R. 1983. "Reply to Shoemaker". In Cohen, L. J. and Hesse, M. (eds). *Aspects of Inductive Logic*. Oxford: Oxford University Press, 313–320.

Andrea Borghini
Recombination for Combinatorialists

Armstrong's combinatorial theory of possibility – or, for short, combinatorialism – owes its name to the fact that claims of possibility and necessity are analyzed in terms of recombinations of pieces of the actual world. The core of the theory is the principle of recombination, a Humean principle endorsed also by David Lewis in order to specify which possibilities can be accounted for within modal realism. While the principle has been extensively discussed in connection to Lewis's metaphysics (e.g. Borghini and Lando 2015; Darby and Watson 2010; Cameron 2008; Efird and Stoneham 2006), far less attention has been devoted to the role of the principle in Armstrong's metaphysics. Following some introductory remarks that provide a context for the discussion (§1), in this paper I first offer a formulation of the principle in keeping with Armstrong's theory (§2-3); hence, I use such formulation to rebut two chief objections to Armstrong's combinatorialism, resting respectively on the mereological structure of entities to be recombined (§4) and on the possibility of scenarios involving alien properties (§5).

1 Introduction

To appreciate the contribution of combinatorialism and to discuss its details, it is important to frame the theory within the broader context of the debate on the nature of possibility and necessity. Combinatorialism aims to provide an answer to the *Problem of Possibility* (PP):

PP: What does it take for a certain situation to be possible?

For instance, what does it take for it to be possible that Foffo the cat will have some fish and potatoes tomorrow? The general nature of the question that sustains the PP suggests that we break it down into three distinct sub-problems, concerning respectively the semantics, the epistemology, and the metaphysics of possibility and necessity; hence *The Semantic Problem of Possibility* (SPP), *The Epistemic Problem of Possibility* (EPP) and *The Metaphysical Problem of Possibility* (MPP):

SPP: What does it mean to say that a certain situation is possible?
EPP: How do we come to know that which is possible?

Andrea Borghini: College of the Holy Cross, Worcester, MA, email: aborghin@holycross.edu

MPP: What sort of entity is a possible entity (of any given kind – a possible individual, property, state of affairs, or …)?

Combinatorialism contributes most directly to the MPP. To the extent that the principle of recombination and the entities that such principle involves are intelligible, combinatorialism suggests a solution also to the EPP. As for the SPP, combinatorialism relies by and large on possible-worlds semantics. In possible-worlds semantics, the truth conditions of propositions – and, in particular, of propositions involving modal terms – are evaluated with respect to a class of possible worlds. A proposition involving a possibility claim is true at a world when true *at a possible world*, while a proposition involving a necessity claim is true at a world when true *at all possible worlds*. For instance, the proposition *Foffo the cat could eat fish and potatoes* is true at our world if there is a possible world where Foffo does eat fish and potatoes; and the proposition *Foffo the cat is necessarily a cat* is true at our world when Foffo is a cat at every possible world. By means of this simple semantic proposal, which can be further articulated in various ways (cfr. Menzel 2015 and Girle 2003), over the past fifty years philosophers have addressed old philosophical questions concerning modality in a new guise (for a historical reconstruction of possible-worlds semantics, see Ballarin 2010 and Copeland 2002). Yet, the proposal raised puzzling questions regarding the metaphysical and the epistemological aspects of possibility: what is a possible world? And how do we know which worlds exist?

The difficulty with the EPP resides in the fact that the majority of possible entities are mere possibilia – they have never been, and will never be, actual, hence we cannot experience them directly. For instance, Foffo could have had, but never did have, a scar on the face. We can know what Foffo looks like by being acquainted with Foffo, or by relying on someone's report who was acquainted with Foffo; but how do we know that Foffo could have had a scar on the face? No one can ever see that scar. Perhaps such possibility is a projection of our minds; or perhaps it rests on an inference based on certain empirical data (Foffo could have had a scar based on the fact that other cats in similar situations did develop scars); or perhaps it is a deduction (Foffo could have had a scar because nothing contradicts our thinking that he does have a scar).

Like other modal theories (e.g. modal realism or abstract ersatzism), combinatorialism addresses the problems posed by the EPP by turning to the MPP for help. The key tenet is that constituents of the actual world provide sufficient ground to accommodate all the truths that a theory of modality must accommodate; they do because each possible world is a *combination* of the entities of the worlds that differs for some respect from the way these entities are actually arranged. Thus, to the extent that we can have knowledge of (i) the constituents of the actual world and of (ii) the ways in which those constituents can be recombined, we can have knowledge of possible and necessary scenarios.

Such a proposal, however, gains credibility only to the extent that we have a principled manner to tell which combinations of constituents of the actual world are really possible. This is the key role, which is assigned to the principle of recombination. In order to introduce the combinatorialist version of the principle of recombination, however, we must first discuss the entities that the principle recombines – the *constituents* of states of affairs. In the next section we shall discuss constituents, to then formulate the principle in the following section.

2 Constituents of states of affairs

A state of affairs is a complex whole made out of various constituents. Consider, for example, the following sentence:

(1)   Foffo eats fish and potatoes.

According to the combinatorialist, (1) expresses the existence of a state of affairs, made out of Foffo, fish and potatoes. These three are the *constituents* of the state of affairs; they are arranged in a specific order and, for this reason, we can claim that such state of affairs is structured. Now, states of affairs need not to be structured wholes; consider for instance the state of affairs expressed by the sentence:

(2)   It rains.

Such state of affairs arguably has one constituent only, namely rain (or raining). Nonetheless, aside for a few instances such as the one illustrated by (2), all states of affairs will have some structure, that is, they will have at least two constituents, related in a specific order and covering some specific roles.

The elements combined by the principle of recombination are the constituents of the states of affairs. It is hence important to see more in details the notion of constituent. For Armstrong, constituents divide into two fundamental categories: individuals (sometimes called also 'particulars' or 'objects') and universals. Foffo is an individual. Being a cat is a universal. Multiple criteria have been proposed to differentiate between individuals and universals, yet we need not concern us here with discussing them. It is important, though, to stress that Armstrong's combinatorialism relies on the distinction between individuals and universals: were we to reject such distinction, we would thereby affect how constituents can recombine.

Even if we take for granted that there is a distinction between individuals and universals, it is not thereby settled how many constituents are involved in each state of affairs. This is because constituents are individuated by means of the logical analysis of the structure of a sentence, and such analysis can be carried out in at least two ways: (i) through the identification of *logically relevant roles* or (ii) through the identification of *logically relevant constituents*.

(i) The first modality of individuation of the constituents of a state of affairs relies on logically relevant roles and proceeds as follows. For any state of affairs, consider a corresponding sentence that expresses that state of affairs. Within such sentence, track which logical roles are occupied (noun, predicate, adverb, etc.) and how many times each role is occupied. Such a modality has important drawbacks. First of all, there is arguably more than one sentence expressing the very same state of affairs; yet, the roles that are occupied may vary from sentence to sentence or may be occupied a different number of times. For example, the following two sentences allegedly express the same state of affairs:

(3) Between Foffo and Fufi there is an age difference.

(4) Foffo is older than Fufi, or Fufi is older than Foffo.

Yet, while in (3) the noun role is occupied three times (by Foffo, Fufi, and age difference), in (4) it is occupied only twice (by Foffo and Fufi). Secondly, the analysis does not dig deep enough into the structure of the sentence. From a combinatorialist point of view, it is key to identify the ultimate elements to recombine, so that no possibility is left overlooked. However, if logical roles are used to individuate the constituents of states of affairs, then many elements will escape the combinatorialist analysis. For instance, the role of the subject is often fulfilled by multiple individuals, so where the analysis in terms of logical role will count one element to recombine, we intuitively will have multiple elements; similar problems arise when we have predicate compounds, and so multiple universals occupying one role.

(ii) For these reasons, the combinatorialist should support the second modality of analysis, based on the concept of the logical constituent of a sentence. The logical constituents in a sentence are the objects, events, and universals that fill in the logical roles within the sentence; thus, in the sentence:

(5) Australian cats are docile.

the logical constituents are each of the Australian cats, which cover the noun role, and the universal docility.

Now, an important distinction is the one between atomic and non-atomic constituents. An atomic individual is such that it has no proper parts, while an atomic universal is such that it is not a conjunction, a disjunction, or a structure made out of other universals. For example, in (5) we may suppose that the atomic constituents are: the predicates 'to be docile' and 'to be Australian'; and each of the cats. On the other hand, in the state of affairs expressed by (3) there is also a non-atomic constituent, namely 'between Foffo and Fufi'. This constituent is not atomic insofar as it is obtained based on other constituents (Foffo, Fufi, some relation expressed by 'between').

Now, the identification of the atomic constituents of reality is not nearly a banal a process. Can a cell be considered an atomic constituent, or is it composed of other constituents? Are colors, sounds or tastes constituents of reality or mere figments of our imaginations? Are gaps and omissions constituents of reality? However, the combinatorialist can partially defuse this issue by pointing out that those are general philosophical problems that do not only surface for the combinatorialist. Following Armstrong, the combinatorialist typically believes that it is not (only) up to philosophers to give an answer to these types of questions. Combinatorialism brings a conceptual clarity to the PP; but, the fact that it cannot have the last word about what things are possible is not strange; to find out, we must also consult other disciplines.

In the simplest model proposed by Armstrong, which shares some important elements also with the metaphysical picture provided by Wittgenstein in the *Tractatus*, all the constituents are atomic (cfr. Armstrong (1997), Skyrms (1981) and Wittgenstein (1921).) Now, imagine having a stock of particulars and universals such that: (i) each particular in the stock instantiates some universal and, furthermore, (ii) each universal in the stock is instantiated by some particulars. Imagine that the world fulfilling such condition is the actual world, so that any recombination of constituents of the actual world delivers a world made out of atomic constituents and that the totality of recombinations associated to the actual world recombines the totality of individuals and universals we have in stock. If this is the case, then the actual world is what Amrstrong calls a *Wittgenstein world*.

As long as combinatorialism assumes that the actual world is a Wittgenstein world, the combinatorial machinery for combinatorialism is relatively straightforward. Unfortunately, such assumption is implausible. For all we know, some important constituents of the actual world (e.g. physical, chemical, or biological kinds) may not be atomic, but rather exhibit a variety of characteristic structures. As we shall see in §4, the possibility of non-Wittgenstein world represents a difficulty for combinatorialism. Before tackling this issue, however it behoove us to provide a combinatorialist formulation of the principle of recombination.

As long as combinatorialism assumes that the actual world is a Wittgenstein world, the combinatorial machinery for combinatorialism is relatively straightforward. Unfortunately, such assumption is implausible. For all we know, some important constituents of the actual world (e.g. physical, chemical, or biological kinds) may not be atomic, but rather exhibit a variety of characteristic structures.

3 Recombination for combinatorialists

Once we are equipped with a stock of individuals and universals, we can provide a full-fledged metaphysical analysis of worlds. The actual world is a vast array of individuals that instantiate a large number of universals. Combinatorialism is the view that the individuals and the universals of the actual world can be rearranged. Let us offer an example. Suppose that, in the actual world, it is the case that:

(6) Foffo eats milk and cookies and Fufi eats fish and potatoes.

Now consider the following sentence, which countenances the same constituents as (6), but recombined in a different manner:

(7) Foffo eats fish and potatoes and Fufi eats milk and cookies.

The core idea of combinatorialism is to regard the scenario depicted in (7) as representing a genuinely possible scenario insofar as it countenances a legitimate recombination of constituents of the actual state of affairs described in (6). Hence, for the combinatorialist, from the fact that (7) is possible we can infer that the following sentence is true:

(8) It is possible that: Foffo eats fish and potatoes and Fufi eats milk and cookies.

The truth conditions for (8) are thus explained by the combinatorialist in terms of the truth conditions for (6). More generally, the core idea of combinatorialism is to regard any representation of a state of affairs, which consists of a legitimate recombination of constituents of the actual world, as a representation of a genuine possibility.

Next on the agenda is the idea of recombination vouchsafed by the combinatorialist, which sets the boundaries of legitimate recombinations. The idea is succinctly captured as follows:

R: A state of affairs S is a recombination of certain states of affairs S* if and only if:

(i) all constituents of S belong to at least one of the states of affairs in S*
or are obtained by interpolation or extrapolation from constituents of
at least one of the states of affairs in S*;
(ii) S is distinct from all the states in S*.

According to R, a state of affairs recombines other states of affairs if and only if it possesses a different order of constituents or if it relates constituents that are not related in the original state. The appeal to interpolation and extrapolation allows some departure from actual states of affairs. A universal is obtained from interpolation of another when some of its aspects are changed; for instance, imagine that the gravitational constant would have some slightly different value than in fact it has. A universal is obtained by extrapolation from another one by removing some of its aspects; for instance, imagine having an electron, with no charge. Part (ii) of the definition, instead, is used to guarantee that S is not actual, but it is a mere possibility, so that we do not call a recombination some states of affairs that in fact we already have in the initial domain of states to be recombined. Thus, suppose the following two sentences represent two actual states of affairs:

(9)   Laura is attracted to Pietro,

(10) Giovanni admires Gina.

Based on R, the following sentence represents a state of affairs that is a genuine possibility:

(11) Laura is attracted to Giovanni.

R licenses also more far fetched possibilities; for instance, suppose that gattraction represents a form of attraction which maintains all the features of attraction in the actual world and, on top of those, it is such that the person who feels the attraction is also pulled by a gravitational force to the person to whom she is attracted. Gattraction is an interpolation of the attraction of the actual world and, as such, it is licensed by R. Hence, the following sentence represents a state of affairs which is:

(12) Laura is gattracted to Giovanni.

However, things aren't quite that simple. In fact, not all recombinations are acceptable. For example:

(13) Giovanni is attracted to admires,

which is a recombination of (9) and (10), is not acceptable. In order to explain which combinations are legitimate, it is necessary to reintroduce the logical analysis based on roles introduced earlier on: (13) would not count as legitimate insofar as 'admires' fulfills the role of a term, while it should fulfill the role of a predicate. Combinatorialism, therefore, is based essentially on a double logical analysis of sentences, which renders a *double metaphysical analysis of states of affairs*: that of atomic constituents and that of atomic roles.

Having looked into some key details of the idea of recombination as understood by combinatorialists, we can now proceed to formulate a principle of recombination that seems suiting for their position:

PRC: Any collection of individuals $x_1,\ldots,x_n$ and universals $U_1,\ldots,U_n$ could coexist, or fail to coexist, with any collection of individuals $y_1,\ldots,y_n$ and universals $V_1,\ldots,V_n$, as long as:
(i) the individuals fit the maximal external relation of a world, and
(ii) the individuals and universals jointly constitute a state of affairs, which is a legitimate recombination of some actual states of affairs.

The principle of recombination plays an important role also in modal realism. It is debated how exactly the principle should be formulated in that context; without entering into details that do not pertain to us here, we can recap the principle as follows:

PRM: For any collection of individuals $x_1, \ldots, x_n$ that stand in a maximal external relation, there is a collection of individuals $y_1, \ldots, y_n$ such that:
(i) each of the $y_1,\ldots,y_n$ is numerically distinct from each of the $x_1,\ldots,x_n$;
(ii) y_1 is an intrinsic duplicate of $x_1,\ldots,y_n$ is an intrinsic duplicate of x_n;
(iii) the $y_1, \ldots, y_n$ compose a world.

There are three traits of PRC that are most notable and that set it apart from PRM. First of all, PRC recombines not only individuals, but also universals; in fact, PRM follows a nominalist metaphysics, where only individuals exist, while PRC rests on a so-called realist metaphysics, where universals – besides individuals – are real too. In general, PRC aims at recombining *all* constituents of states of affairs. The second difference between PRC and PRM is PRC's requirement that the recombination delivers a state of affairs. The third is that the state of affairs must be a legitimate recombination of some actual state of affairs, while PRM accords no special role to the actual world (after all, for Lewis, "actual" is an indexical term and carries no special metaphysical status).

Furthermore, PRC is exempt from some of the difficulties proper of PRM. This is because PRM is formulated in terms of counterparts; yet, since the combinatorialist admits that there is identity among individuals belonging to distinct worlds (after all, the individuals of other worlds exist as surrogates of individuals of the actual world), she will recombine exactly the individuals of the actual world.

Now, Armstrong contends that combinatorialism is capable to provide a reductive analysis of modality, on a par with modal realism. The reason is that combinatorialism explains the meaning of modal sentences in terms of the existence of non-modal, actual entities. Constituents of the actual world provide sufficient ground to accommodate all the truths that a theory of modality must accommodate. However, contra Armstrong's contention, we can point out that the appeal to *legitimate* recombinations makes PRC a circular definition of the concept of possibility: possible states of affairs are those that are *possible* within the limits of the assigned metaphysical roles. Surely, this is a theoretical limitation of combinatorialism in comparison to modal realism; yet, if we consider that modal realism is the only theory that claims to be able to define the concept of possibility in a non-circular manner, the concerns raised by the limitation are minor. So, the combinatorialist offers us an explanation of the concept of possibility, but not a definition. And this may be considered sufficient: in order to offer an adequate solution to the PP, EPP and MPP, it is not necessary to eliminate the concept of possibility; it need only be explained.

PRC provides us with some insight into the combinatorialist solutions of the MPP and the EPP. As for the EPP, combinatorialism seems to play upon two distinct types of knowledge. On one hand, we have the principle of recombination, which states that all legitimate recombinations are possible. It is a metaphysical principle, whose knowledge is gained (to a certain extent) independently of experience. On the other hand, we have the empirical knowledge of the atomic constituents of reality: it is based on this knowledge that we can grasp the meaning of any modal sentence.

As for the MPP, we can sum up the combinatorialist solution as follows:

C: A state of affairs is possible if and only if it is obtained by means of a legitimate recombination of constituents of at least one actual state of affairs.

Because worlds are ways in which elements of the actual world could be recombined, we can claim that for combinatorialism worlds are surrogate of the actual world and, thus, that combinatorialism is an *ersatzist* theory.

4 Rebutting the trickle-down objection

We shall now address a serious worry for combinatorialist. Armstrong (2004) argues that combinatorialism may sit best with atomic worlds. And yet, it is far from an established (scientific or philosophical) fact that our world is atomic. So, the combinatorialist cannot assume that our world is a Wittgenstein world and she must provide a criterion for recombining structured constituents. So, suppose that, among actual states of affairs, we find some structured constituents. Should their parts be recombined as well? For instance, consider a molecule of water and suppose to recombine the molecule: should its atoms of hydrogen and oxygen be recombined as well, or do they not trickle down in the recombination? Call this the *trickle-down objection*, which was first discussed by Sider (2005). Sider suggests one way to address the objection, which nonetheless risks of restricting the combinatorial power of PRC and, hence, to restrict the domain of possible scenarios countenanced by combinatorialism. I shall first present Sider's solution, and then suggest an improvement.

When facing the trickle-down objection, for each constituent in the basis of recombination (the pool of constituents that are suitable to be recombined), the combinatorialist has two sorts of options: either all the structural features of the constituent do trickle down and the result is a recombinant that is as structurally complex as the recombining constituent; or, the structural features do not trickle down, so that the result is an atomic recombinant. (There is also an intermediate option, which is forgone by Sider, according to which some of the structural features of the constituent do trickle down, and some do not, so that the result is a recombinant that retains only some of the structural complexity of the recombining constituent. We shall leave this option on a side for the moment and come back to it later.)

Now, Sider argues that both options are problematic, when applied to all individuals and all properties in the basis of recombination. If none of the features trickle down, then PRC delivers only mereologically simple individuals and unstructured universals, that is, PRC delivers only Wittgenstein worlds. This seems too strict of a limitation of the range of possibilities: suppose we agree that *Being a cell* is a structural universal; then the PRC would obliterate the (merely) possible existence of cells and, hence, PRC would suggest that cells could not have been otherwise. More generally, if none of the features trickle down, then no structured individual and no structural universal could have been instantiated otherwise.

On the other hand, if all the structural features of constituents trickle down, then PRC delivers some puzzling worlds: consider, for instance, recombining an individual that is mereologically simple (and thus instantiating a non-structural universal in the actual world) with a structural universal; if all the features of the structural universal trickle down, then the recombination delivers a world where a mereological simple instantiates a structural universal – a metaphysical monster, or a metaphysical impossibility, if one maintains that structural universals can be instantiated only by individuals whose mereological structure matches the structure of the universal.

A first intuitive solution of the dilemma – the one suggested also by Sider – is to reserve different treatments for individuals and for universals. All the features of universals should trickle down, while none of the features of individuals should trickle down. The recombination of a structural universal whose structure involves – say – twenty-seven parts, would hence involve twenty-seven individuals, one for each part of the structure.

The solution brings to light a general limitation of the various principles of recombination so far proposed, including PRC and PRM: they are blind to the mereological structure of the individuals in the basis of recombination (on this point, see Borghini and Lando 2015). The solution, however, does little to solve the limitation. First of all, within the solution the mereological structure of the recombined individuals is derivative upon the structures of the universals that the individuals instantiate; however, the structures of the universals arguably followed the mereological structure of some individuals; hence, the mereological structure of the recombined individuals seems to emerge magically, out of thin air. Secondly, the solution fails to deliver gunky scenarios, unless one postulates the existence of gunky universals (cfr. Borghini and Lando 2015 on this point). It is controversial whether gunk is a genuine possibility; but, if one agrees that it is, the only way in which the solution could guarantee the existence of gunk is to postulate the existence of some structural universal with an infinitely complex structure – a postulation that would be rejected by most metaphysicians.

A more palatable solution rests on the intermediate option suggested above, according to which *some* of the structural features of the constituent do trickle down, and some do not. Being selective with respect to the features that trickle down is entirely in keeping with the spirit of PRC: the selective removal of some mereological features of individuals or of some structural features of universals is indeed a special case of extrapolation, which is part and parcel of the combinatorial theory of possibility. Hence, I suggest that the best interpretation of PRC is one where the principle licenses as many features to trickle down as possible, so long as no puzzling states of affairs are generated.

It may be objected that the limitation of the extent of features of constituents that trickle down is a limitation of the theoretical power of combinatorialism, in that the theory fails to deliver modal plenitude. Yet, the combinatorialist will insist that the selective obliteration of features ensures that only those features that would deliver metaphysically impossible scenarios are hindered from being recombined; hence, modal metaphysical plenitude is attained by PRC, although modal logical plenitude is not. A Humean may be dissatisfied with this answer, as for the Humean anything can coexist, or fail to coexist, with anything, hence all constituents – with no restrictions with respect to their features – should be included in the basis of recombination.

5 Rebutting the objection from alien possibilities

The second objection to the PRC that requires attention is related to ways in which the actual world might not contain adequate ontological resources for creating a basis of recombination that accommodates our intuitions about modality. In particular, it concerns the possibility that the actual world may have contained either more individuals, or more universals. Let us discuss these options, starting from the one regarding individuals.

To begin, consider the possibility that the actual world could have contained fewer individuals; this scenario is typically referred to as a contraction. The combinatorialist definitely has adequate resources to represent a contraction: all she needs is a world made out of a conjunction of states of affairs that jointly contain less individuals than the ones existing in the actual world. It is a different story, instead, when we consider the case of an expansion, that is, of a world that contains more individuals than in the actual world. Suppose the stock of recombinable individuals from the actual world includes n individuals, and we are given the sentence:

(14)   There could have been $n+1$ cats.

Regardless of whether (14) would turn out to be true or false, (14) seems to express a sentence whose content is straightforward. So, the combinatorialist can hardly appeal to the fact that, despite our modal intuitions, (14) expresses an impossible state of affairs. Yet, how can her theory represent such a state of affairs, if there are not enough individuals in the actual world to deliver it?

A solution may be to add a clause regarding the legitimacy of recombinations obtained through the iteration of elements of the actual world: if in the actual world there are n individuals, a state of affairs with $n+1$ individuals is possible insofar as it is derived from the actual world by the iteration of an individual constituent. Thus, a world in which there are more individuals than there are in the actual world becomes a possibility by iterating some items of the actual world.

Whether iteration can be regarded as a legitimate operation, from a metaphysical point of view, may nonetheless be doubted. One way to understand iteration is by way of fictionalism: iteration is a pretense; for instance, it's the pretense that, instead of there being only one pope, we have two (or three or ... infinitely many) of them. Modal fictionalism, however, carries an ontological commitment that is typically orthogonal with respect to the combinatorialist commitment. Indeed, the earliest formulation of modal fictionalism (namely Rosen 1990; cfr. also Nolan 2007) built upon modal realism; it is dubious that the basis for recombination of PRC could contain both constituents modeled after an Armstrongian ontology and constituents modeled after a Lewisian ontology. More importantly, modal fictionalism typically buys a full theoretical package – e.g. the whole modal realism – and it is dubious how such package could be made compatible with the combinatorialist perspective. Of course, one could devise a more suitable version of modal fictionalism, with a fiction based on combinatorialism rather than on modal realism; still, fictional entities are ontologically spurious: a state of affairs whose constituents are partially fictional and partially non fictional has little metaphysical plausibility and the combinatorialist would have to provide much more context for such sort of entity in her ontology.

More promising may be some sort of ersatzist understanding of iteration, according to which iterated items are surrogates of actual items. The surrogates could be linguistic entities, or abstract entities, or images. Thus, if the actual world has n individuals, there could be a world with $n+1$ individuals because in the basis of recombination we find not only concrete individuals but also some words/abstract entities/images that play the role of individuals. Yet, also in this case, the objection remains that the basis of recombination would contain a heterogeneous array of individuals, some concrete and some surrogates of some of the concrete: the metaphysical plausibility of a state of affairs partially concrete and partially ersatz remains to be substantiated.

A more promising strategy to address cases of expansion in a combinatorialist fashion is to treat them as cases of interpolation, where the feature of individuals that is altered is mereological complexity. It is possible to generate a world with more individuals than our world by interpolating the mereological complexity of our world (taken as a large mereological sum of individuals) and making such structure even more complex. Whether this solution is viable, however, depends on how interpolation is understood. In particular, the question is whether it is legitimate to expect that by means of interpolation we can augment the ontological resources of the actual world. Armstrong himself does not say much about this operation, so it is difficult to settle the matter at this time. The strategy on interpolation, at any rate, seems the most promising.

Let us now move to consider the second typology of possibility that the combinatorialist does not seem to be able to explain: the case of alien universals. Not only can we think that there are worlds with more individuals than there are in our own; there could also be worlds with individuals whose qualities and abilities are completely distinct from those of the individuals of our world. Once, these types of possibilities were excluded by invoking the so-called *principle of plenitude* for the actual world, according to which God created (in the actual world) everything God could have created, not only by number (which is debatable, but less relevant), but also, and more importantly, by variety. The fact remains that this principle does not seem to have a clear empirical confirmation: there are undoubtedly many species of living organisms; but, why should we believe that there couldn't be or couldn't have been even more? From the record we have, indeed, there are many regions of the morphospace of organisms that are left unoccupied. Or why couldn't the periodic table, for example, have included more elements than the ones it already includes? In other words: why believe that we live in the most variegated world that could exist?

The appeal to extrapolation and interpolation in the PRC is precisely to be able to meet some of these alien scenarios. The extent of scenarios that we can meet, though, depends on how the two operations are understood. If during the operations we can only modify a universal by removing some of its features, or by adding features that belong to other universals that are actually instantiated, then any feature that is alien to our world cannot be licensed through the PRC and, therefore, any state of affairs involving alien features is – according to the combinatorialist – impossible.

Perhaps the combinatorialist should endorse a bolder understanding of extrapolation and interpolation, according to which the operations can make use also of universals that are actually uninstantiated. This solution, though, would require an important change in the conception of universals; it would require, that is, that there are some universals that are not actually instantiated.

6 Conclusions

The principle of recombination plays a key role in Armstrong's metaphysics. Despite its importance, the principle has received little attention so far in the literature on combinatorialism. Drawing also on the parallel work on the principle of recombination in modal realism, in the present work I aimed to clarify a formulation of the principle and to defend such formulation against two important objections: the so-called trickle down objection and the objection moving from the possibility of alien individuals or universals. As we have seen, combinatorialism can accommodate the trickle down objection by appealing to an intermediate option, according to which only selected structural features of a constituent trickle down, so that all and only the features that are not metaphysically puzzling are selected. On the other hand, the solution of the objection moving from the possibility of alien individuals or universals is more costly. Combinatorialism can countenance the possibility of worlds containing alien individuals or universals only by endorsing a bold understanding of extrapolation and interpolation, at the cost of revising – among others – some important tenets of the theory of universals; whether the combinatorialist can and should follow this strategy remains to be assessed.

Bibliography

Armstrong, D. M. 1978. *Nominalism and Realism (2 voll.)*. Cambridge: Cambridge University Press.
Armstrong, D. M. 1986. "The Nature of Possibility". *Canadian Journal of Philosophy*, 16, 575–594.
Armstrong, D. M. 1989. *A Combinatorial Theory of Possibility*. Cambridge: Cambridge University Press.
Armstrong, D. M. 1997. *A World of States of Affairs*. Cambridge: Cambridge University Press.
Armstrong, D. M. 2004. "Theorie Combinatoire Revue et Corrigeée". In Monnoyer, J.M. (ed.), *La Structure du Mond: Objets, Propriétés, États et Choses*. Paris: Vrin, 185–198.
Ballarin, R. 2010. "Modern Origins of Modal Logic". In Zalta, E. N. (ed.), *The Stanford Encyclopedia of Philosophy* (Winter 2014 Edition), URL = <http://plato.stanford.edu/archives/win2014/entries/logic-modal-origins>.
Borghini, A. and Lando, G. (2015), "Natural Properties and Atomicity in Modal Realism". *Metaphysica*, 16, 103–121.
Bradley, R. 1989. "Possibility and Combinatorialism: Wittgenstein Versus Armstrong". *Canadian Journal of Philosophy*, 19: 15–41.
Cameron, R. 2008. "Recombination and Intrinsicality". *Ratio*, 21, 1–12.
Copeland, B. J. 2002. "The Genesis of Possibile Worlds Semantics". *Journal of Philosophical Logic*, 31, 99–137.
Darby, G. and Watson, D. 2010. "Lewis's Principle of Recombination: Reply to Efird and Stoneham". *Dialectica*, 64, 435–445.
Efird, D. and Stoneham, T. (2006), "Combinatorialism and the Possibility of Nothing". *Australasian Journal of Philosophy*, 84, 269–280.
Girle, R. (2003). *Possible Worlds*. Montreal: McGill–Queens University Press.
Kim, S. 1986. "Possible Worlds and Armstrong's Combinatorialism". *Canadian Journal of Philosophy*, 16, 595–612.
Lewis, D. K. 1992. "Armstrong on Combinatorial Possibility". *Australasian Journal of Philosophy*, 70, 211–224.

Menzel, C. 2015. "Possible Worlds". In Zalta, E. N. (ed.), *The Stanford Encyclopedia of Philosophy* (Spring 2015 Edition), URL = <http://plato.stanford.edu/archives/spr2015/entries/possible-worlds>.

Nolan, D. 2007. "Modal Fictionalism". In Zalta, E. N. (ed.), *The Stanford Encyclopedia of Philosophy* (Winter 2007 Edition), URL = <http://plato.stanford.edu/archives/win2007/entries/fictionalism-modal>.

Rosen, G. 1990. "Modal Fictionalism". *Mind*, 99, 327–354.

Thomas, H. G. (1996), "Combinatorialism and Primitive Modality". *Philosophical Studies*, 83, 231–252.

Sider, T. 2005. "Another Look at Armstrong's Combinatorialism". *Noûs*, 39, 680–696.

Skyrms, B. 1981. "Tractarian Nominalism". *Philosophical Studies*, 40, 199–206.

Wang, J. 2013. "From Combinatorialism to Primitivism". *Australasian Journal of Philosophy*, 91, 535–554.

Wittgenstein, L. 1921. "Logisch–philosophische Abhandlung". In *Annalen der Naturphilosophie*, 14, 185–262; revised English edition, *Tractatus Logico–Philosophicus*. London: Kegan Paul, Trench, Trubner & Co. 1922.

Michele Paolini Paoletti
Who's Afraid of Non-Existent Manifestations?

1 Introduction

According to many recent metaphysical accounts of powers,[1] powers are fundamental, irreducible entities.[2] Even if you do not believe that powers are the *only* sort of fundamental entities – or the only sort of fundamental properties –, as some philosophers such as Alexander Bird (2007) maintain, you could still believe that they are part of the basic ontological inventory of the universe, that God could not produce an exhaustive copy of our universe without reproducing powers and their instantiations.

Moreover, it seems that powers are essentially individuated – among other – by their (possible) manifestations. Thus, the power to dissolve salt – that is seemingly possessed by water – is also essentially individuated by its (possible) manifestation, i.e., dissolving salt. That power is what it is – the power to dissolve salt – also in virtue of that (possible) manifestation. If it had had another (possible) manifestation, it would not have been that power, but the power to something else. If it had had no (possible) manifestation at all, it would not have been a power at all. Of course, some powers are associated with different (possible) manifestations in different (possible) circumstances. Yet, this does not imply that they are not essentially individuated also by those (possible) manifestations: perhaps, they are individuated by all those (possible) manifestations and all those (possible) circumstances.

Anyway, I shall argue in this paper that, if you claim that powers are fundamental, irreducible entities and that they are essentially individuated also by their (possible) manifestations, you should also claim that there are non-existent objects.[3] In this respect, I shall agree with David M. Armstrong on the thesis that a powers metaphysics is committed to the truth of Meinongianism, i.e., the doctrine

---

[1] I take "powers" here as referring to powers, dispositions, capacities and propensities, because the distinctions between these sorts of entities are not relevant in this paper. See Choi, Fara 2012.
[2] See, for example, Mumford 1998, 2004, Cartwright 1999, Ellis 2001, Molnar 2003, Bird 2007, Martin 2008, Mumford and Anjum 2011.
[3] See, for example, Armstrong 1997, 79.

Michele Paolini Paoletti: University of Macerata, email: michele.paolinip@gmail.com

according to which there are non-existent objects. However, I shall disagree with him on the evaluation of this fact, since I do not believe that this constitutes a problem at all.

In section 2 of this paper, I shall introduce my argument. In section 3, I shall consider Armstrong's analysis of seemingly true ascriptions of powers. In section 4, I shall examine five alternatives that are not committed to the truth of Meinongianism and, in section 5, I shall deal with some miscellaneous concerns about Meinongianism.

2 The argument

Ascriptions of powers are often thought of as what is expressed by seemingly true statements such as

(1)   I can raise my arm (i.e., I have the power to raise my arm).

If you think that powers are part of the basic ontological inventory of the universe and if you accept that my power to raise my arm cannot be reduced to other powers (possessed by me or by other entities), you are committed to the existence of my power to raise my arm or, more generally, of the power to raise one's arm. Moreover, powers are often considered properties – particular or universal. Thus, if I have the power to raise my arm, it is either the case that there is a particular property such as my power to raise my arm – that characterizes or constitutes me[4] – or that there is a universal property such the power to raise one's arm – that is instantiated by me. The possession of a power does not imply its activation. It could be true that (1) and false that

(2)   I raise my arm.

In principle, powers could remain unmanifested. Finally, I assume here that all powers are essentially individuated also by their (possible) manifestations. Together with the thesis that some powers could remain unmanifested, this implies that, nevertheless, those powers are essentially individuated by their merely pos-

---

[4] A particular property *characterizes* something iff it is one of its modes (see Lowe (2006)). Modes ontologically depend on their "bearers". On the other hand, a particular property *constitutes* something iff it is one of the tropes that constitute that thing. In fact, tropes are particular properties that are more fundamental than ordinary objects and that constitute such objects in appropriate conditions (see Maurin (2013)).

sible manifestations, i.e., by manifestations that never exist or occur, even if they could have existed or occurred. The move towards Meinongianism is quite easy.

I shall consider one particular case: the case of generative powers. Generative powers are powers to generate something. For example, a certain existing cell (let me name it "Cell-1") has the generative power to generate by mitosis two different cells (let me call them "Cell-2" and "Cell-3"). Perhaps, this power is never activated. Yet, Cell-1 could still possess this power, even without its activation. Of course, Cell-2 and Cell-3 cannot exist before the activation of that power, nor can they come into existence if that power is not activated. My argument runs as follows:

(a) generative powers are fundamental, irreducible entities;
(b) if it is true that (a), then there are non-existent objects and such objects are fundamental;
(c) thus: there are non-existent objects and such objects are fundamental (from (a) and (b), by MP).

I shall prove (a) dialectically in the next sections. However, let me consider the case in which Cell-1 generates Cell-2. If Cell-1 generates Cell-2, then it was at least metaphysically possible that Cell-1 generated Cell-2.[5] If it was at least metaphysically possible that Cell-1 generated Cell-2, then it was either the case that Cell-1 had the power to generate Cell-2, or that it had the power to acquire the power to generate Cell-2 in relevant circumstances.[6] Thus, Cell-1 either had the power to generate Cell-2, or it had the power to acquire the power to generate Cell-2. In both cases, there was a power essentially involving a reference to Cell-2, even before Cell-2's starting to exist. Moreover, it seems that Cell-1 actually had the power to generate Cell-2: that power only needed to be activated. Thus, there was a generative power, such as the power to generate Cell-2. Is this power ontologically fundamental? Can it be eliminated or reduced to anything else? I do not think that it can. In fact, as I shall argue in the next sections, such eliminations or reductions are inefficacious, since they do not preserve the truth of some ascriptions of powers across actual and possible situations or they do not succeed in getting rid of non-existent objects, or they commit us to entities that are more problematic than non-existent objects.

The truth of premise (b) can be demonstrated as follows. I shall work under the hypothesis that generative powers are fundamental, irreducible entities, i.e.,

---

[5] Of course, this metaphysical possibility could still be reduced to something else, not involving Cell-2, by those who claim that there are no non-existent objects such as Cell-2 before their coming to existence.
[6] See Molnar 2003, 100-101.

the antecedent of the conditional that I aim at demonstrating. Powers are essentially individuated by their (possible) manifestations and – in turn – such manifestations need to be individuated. The existence of something that still does not exist[7] – such as the existence of Cell-2 – is the (possible) manifestation of generative powers (such as the power to generate Cell-2). Yet, this (possible) manifestation is individuated only if something that still does not exist is individuated. Thus, generative powers are individuated only if their (possible) manifestations are individuated and such manifestations are individuated only if something that still does not exist is individuated. Thus, if generative powers are fundamental and irreducible entities (that obviously need to be individuated), there is something that still does not exist and that nevertheless contributes to the individuation of generative powers.

Moreover, if generative powers are fundamental, irreducible entities, non-existent objects are fundamental too. In fact, it seems reasonable to claim that whatever contributes to the individuation of fundamental entities is fundamental too. Generative powers are fundamental entities. Thus, non-existent objects are fundamental too. Thus, if generative powers are fundamental, irreducible entities, there are non-existent objects and such objects are fundamental too.

I anticipate an objection here: if generative powers are fundamental entities, can they depend for their individuation on anything else, i.e., on non-existent objects? Either they depend on something else for their individuation, or they are fundamental, and there is nothing which is both fundamental and depends on something else for its individuation. Yet, I think that one could still maintain that there are certain sorts of fundamental entities that also, but not only depend on something else for their individuation. Of course, if God aimed at creating an exhaustive copy of our universe, He would have to copy those entities too – not only the entities on which they partially depend for their individuation. Powers belong to this category of fundamental entities. In fact, they also, but not only depend for their individuation on their (possible) manifestations – at least if we assume that being a power is a primitive and irreducible feature of powers. Thus, the generative power to generate Cell-2 also, but not only depends on Cell-2 for its individuation. In addition, its being a power does not depend on anything else.

My argument can be criticized in four major ways. Either you demonstrate that (i) generative powers are not fundamental entities, or that (ii) they are fundamental entities which do not depend on non-existent objects for their individuation, or that (iii) those non-existent objects on which they depend for their individuation wholly depend in turn on existent objects or properties, or that (iv) those

---

[7] And that perhaps will never exist.

non-existent objects actually exist. Yet, before turning to these criticisms, I shall consider Armstrong's attempt to get rid of powers *qua* fundamental entities.

3 Armstrong vs. powers

Armstrong's concerns on the fundamentality of powers are also based on his refusal on Meinongianism. Anyway, he thinks that powers ascriptions in general should be reduced to something else. In other terms, powers ascriptions should be analysed into ascriptions of something else to objects and/or to properties. Here is the form of such an analysis:

(powers) as a matter of metaphysical necessity, for every object, that object has a certain power p_1 iff ϕ,

where "ϕ" should substituted by the *analysans* – that does not have to mention powers. The left side of the equivalence ("that object has a certain power p_1") is the *analysandum*. Metaphysical necessity is invoked in order to distinguish appropriate analyses of powers ascriptions from accidental regularities.

Let me now assume that "p_1" stands for the power to produce an instantiation of a certain property H (e.g., the property of being identical with Cell-2). Thus, "P_H" will stand for the property of having the power to produce an instance of a certain property H. If Cell-1 does not actually produce Cell-2, such a power is unmanifested. Armstrong's analysis primarily deals with unmanifested powers. In fact, unmanifested powers are more problematic than manifested ones, since they seemingly introduce in the realm of existence mere possibilities. On the other hand, manifested powers "point towards" existent manifestations: they are not fundamental entities and their analysis can be easily performed in terms of existent entities.

Moreover, I shall assume that "F" is a variable ranging over properties that are not powers and that "constitute" the microstructures of objects having p_1 – as long as their microstructures reveal what those objects are. "G" is a variable ranging over properties whose instantiation – together with the absence of the instantiation of other properties, at least in some cases – is nomologically sufficient to produce the instantiation of H. "J", "K", etc. stand for the properties whose instantiation should be excluded in order for G to be nomologically sufficient to produce the instantiation of H. "N" stands for the relation of nomological necessitation between properties. Finally, I shall assume that the variable "x" ranges over existing objects, "$\Diamond_N$" and "$\Box_M$" are two modal operators, that respectively represent nomological possibility ("given certain laws of nature, it is possible that") and the aforementioned metaphysical necessity. Armstrong (1997: 81-82) offers this analysis of powers ascriptions:[8]

(arm.powers) $\Box_M \forall x(P_H x \leftrightarrow \exists F(Fx \& \exists G N(F \& G(\& \neg J \& \neg K \& \ldots))H \& \Diamond_N Gx))$

Informally: as a matter of metaphysical necessity, any object has the power to produce an instance of H iff there is a microstructural property F instantiated by that object and F and G (and, in case, the absence of J, of K, etc.) nomologically necessitate H and it is nomologically possible that that object instantiates G. The first conjunct of the analysans is the instantiation of a microstructural property, the second is a certain law of nature (laws of natures are relations between universals, from Armstrong's perspective[9]), the third is the nomological possibility of G's instantiation.

Here are some problems with this analysis.[10] Firstly, it seems that some powers can be associated with different microstructural properties: fragility, for example, is realized by vases, glasses, and so on, i.e., by objects having different microstructures. Thus, either one substitutes "F" with a disjunction of microstructural properties or s/he claims that there is a different power for each microstructural property. Yet, in the former case, disjunctions (or disjunctive properties) would turn out to be fundamental (or more fundamental than powers ascriptions), while, in the latter case, one should abandon the idea that there is a universal power ascription – even if it seems that all the objects that have p_1 with different microstructures have the same power.

Secondly, different properties G could be associated with one and the same power. Moreover, such properties could in turn be associated with different negative clauses, excluding the instantiation of certain properties. Thus, the law of nature in the second conjunct could become much more complex, including further conjunctions and disjunctions of properties: $N(F \& ((G_1 \& \neg J) \lor (G_2 \& \neg K) \lor \ldots))H$. Thirdly, negative clauses are problematic, since they are identical with negations of properties. Yet, what is the negation of a property? Within a law of nature, the

---

[8] In this text, Armstrong does not actually talk of microstructural properties. This terminology is introduced, for example, in Armstrong, Martin, Place 2002, 41, by preserving the analysis of unmanifested powers suggested five years before. Moreover, Armstrong (in Armstrong, Martin, Place 2002, 39) affirms that a disposition is a microstructure picked out via its causal role.

[9] See Armstrong 1982. Moreover, Armstrong 2005 claims that laws of nature are relations between types of states of affairs. Anyway, for the sake of simplicity, I shall maintain here that they are relations between universals, since types of states of affairs are, in turn, universals.

[10] For some of the difficulties that I shall briefly examine here, see Bird 2007, 18-42.

negation of a property cannot be the non-instantiation of that property in a certain situation, since laws of nature are relations between properties, and not between instances of properties. Thus, either the negation of a property is a negative property of a property (e.g., the property J is such that it is not instantiated) or it is a negative property (e.g., the property non-J). As it is well known, both disjunctive and negative properties are problematic entities for Armstrong: they do not exist or, at best, they are not fundamental.[11] Thus, a different analysis of powers ascriptions should be provided without invoking them.

However, even if we were inclined to accept such properties, some difficulties would still remain. First of all, there are some powers (finkish powers[12]) that can be lost by objects *after* the instantiation of the relevant G, so that H is not instantiated (i.e., the manifestation does not occur). Such powers should be analysed by including within the relevant law of nature in the second conjunct of the *analysans* the negations of all the conditions that could produce their loss. This means that further negations of properties– constituting further negative clauses – should be added.[13] Yet, such negative clauses are introduced *only because* their corresponding positive properties produce the loss of the finkish power. Thus, each negative clause is there only because it is compatible with the existence *of p_1* after the instantiation of G and before the instantiation of H, while the corresponding positive property is such that is incompatible with the existence *of p_1* after the instantiation of G. This explanation of negative clauses provides a bad analysis for powers ascriptions, since it reintroduces powers in the *analysans* and it makes it the case that the *analysans* is what it is only in virtue of certain facts involving the powers to be analysed.

Secondly, there is a more general problem with conjunctive properties. Armstrong (1978, 30-42) accepts conjunctive universals within his ontology. Thus, the first *relatum* of the nomological necessitation relation in the second conjunct of the *analysans* is a really complex conjunctive property. Yet, it is either the case that *every* conjunction of properties gives rise to a complex conjunctive property that should be included within our ontology and/or that should be invoked as a fundamental entity, or that *only some* conjunctions of properties give rise to complex conjunctive properties to be included within our ontology and/or to be invoked as fundamental entities. However, accepting the first horn of this dilemma, too many complex conjunctive properties turn out to be included within our ontology

---

11 See Armstrong 1978, 19-29 and 2004, 54-67.
12 See Martin 2008, 12-23.
13 Thus, there will be two different kinds of negative clauses: those that only prevent the non-obtaining of the manifestation and those that prevent the non-obtaining of the manifestation by preventing the loss of the relevant power after the instantiation of G.

and/or to be fundamental, and such properties cannot be dispensed with by only accepting their conjuncts. In fact, only conjunctive properties stand in the nomological necessitation relation with H – and not their conjuncts. Yet, following the second horn of the dilemma, one still has to explain why certain conjunctions of properties give rise to conjunctive properties that figure in laws of nature about powers ascriptions, while other conjunctions of properties do not give rise to conjunctive properties. This explanation cannot mention p_1. Yet, it seems that only p_1 provides an adequate explanation: complex conjunctive properties are there *only because* their conjuncts – *put together* – are somehow associated with p_1 and, more precisely, with the activation of p_1. Of course, one could accept the second horn and give no explanation: it is a primitive truth that only some conjunctions of properties give rise to conjunctive properties. Yet, this would imply the substitution of something (a power) with something else which is more complex and perhaps problematic (a conjunctive property) within one's ontology and/or at the fundamental level of the universe.

In sum, Armstrong's project – as it is expressed by (arm.powers) – seemingly fails. Perhaps, there is something unanalysable about powers ascriptions, so that powers should be part of our ontological inventory and/or of our fundamental ontological inventory.[14]

4 Getting rid of non-existent manifestations

I shall now analyse five attempts to get rid of non-existent objects as manifestations (or as part of the manifestations) of generative powers. Such attempts are based on four strategies, that I have mentioned in section 2, even if there is no strict correspondence between attempts and strategies: (i) generative powers are not fundamental entities; (ii) they are fundamental entities which do not depend on non-existent objects for their individuation; (iii) those non-existent objects on which they depend for their individuation wholly depend in turn on existent objects or properties; (iv) those non-existent objects actually exist.

Stephen Mumford (2004: 194-195) claims that powers are directed towards the manifestation of certain universal properties and they do not require particulars within their essences. Thus, the fundamentality of generative powers would not

---

[14] Armstrong 2004, 137-138 gives an analysis of powers ascriptions in terms of conditionals and counterfactuals. Anyway, that analysis seems to be affected by the some problems that I have examined here.

require non-existent objects as fundamental entities. He adds in the same place that

> first, a power is not typically a power to manifest a universal in some very precise way, at a precise time and place. A power might be a power to dissolve, *when and wherever*, and rarely a power to dissolve at spatiotemporal location $p_1 t_1$. Second, because a universal is fully present in its instances, we can note that the thing to which the power is directed will indeed be present whenever we have an actual and specific instantiation. Our universal F is present in the specific manifestation $F(p_1 t_1)$ and is the part of $F(p_1 t_1)$ for which the power was a power.

Mumford suggests that, if powers were powers to produce certain particulars, then they would be essentially individuated not only by that particular, but also by certain spatiotemporal locations. He seemingly thinks of something similar to Kimian events as the particular manifestations of such powers. In fact, Kimian events are essentially individuated by the objects and the *n*-adic properties that are involved in those events and by the time at which they occur. Anyway, one could reply that generative powers are directed either to non-existent objects, or to facts involving non-existent objects (e.g., the fact that a certain non-existent object starts to exist). Non-existent objects and facts are such that they are not essentially individuated by certain spatio-temporal locations: it is metaphysically contingent that a certain object starts to exist at a certain spatio-temporal location and whatever is metaphysically contingent for something is not part of its essence.

The second remark suggests that universals are more fundamental than particular manifestations, since those manifestations exist only because universals are instantiated. Besides reaffirming that it is not necessary that manifestations are essentially individuated by spatio-temporal locations, it is worth noticing that universals are instantiated only because they are instantiated by objects, so that objects turn out to be more fundamental than universals' instantiations.

Strategies (i)-(iii) are compatible with the following reduction scheme of the power to produce Cell-2, which is a paradigmatic case of generative power:

(gen.powers) as a matter of metaphysical necessity, for every object, that object has the power to produce Cell-2 iff Ψ,

where "Ψ" can be substituted by other powers that are not essentially individuated by non-existent objects or by entities that are not powers. Anyway, I have already criticized Armstrong's reduction of powers to other entities. It is now time to examine the former alternative.

At first, one could consider powers towards the instantiation of properties, i.e., powers towards the fact that a certain property P is instantiated – or that it starts to exist, or that it acquires some feature (such alternatives are introduced in order to deal with properties that are not universal, as we will see). Here are some possibilities:

(Ψ1) Ψ = that object has the power to produce an instantiation of the property of being identical with Cell-2;

(Ψ2) Ψ = that object has the power to produce an instantiation of a conjunctive property P_C (that uniquely individuates Cell-2)[15];

(Ψ3) Ψ = that object has the power to produce the existence of a certain mode (i.e., of a certain particular property, that essentially depends on its "bearer"[16]) such as Cell-2's existence or Cell-2's being a cell or whatever else;

(Ψ4) Ψ = that object has the power to produce the existence of a certain aggregation of tropes (i.e., of a certain aggregation of particular properties that do not essentially depend on their "bearers");

(Ψ5) Ψ = there is a certain internal relation[17] between the Platonic universal of having that power and some other Platonic Universal U_C – that presumably uniquely individuates Cell-2.[18]

Following (Ψ1)-(Ψ5), one could either claim that (i) generative powers are not fundamental entities, since they are reduced to other powers, or that (ii) they are fundamental entities which do not depend on non-existent objects for their individuation, but only on existent properties, or that (iii) those non-existent objects on which they depend for their individuation wholly depend in turn on existent objects or properties, so that Cell-2 depends, for example, on tropes – at least according to (Ψ4).

(Ψ1) is hardly acceptable. In fact, it seems that the property of being identical with Cell-2 is essentially individuated by Cell-2, and not the opposite, at least if you do not wish to claim that there are haecceities that are more fundamental than objects – provided that they ground their individuation. (Ψ3) should be dismissed for similar reasons, since it explicitly claims that the mode which starts to exist is

---

15 Armstrong 1995, 619 accepts that non-existents can be reduced to combinations of properties.
16 See Lowe 2006.
17 An internal relation is a relation that wholly depends on the existence and/or on the essence and/or on the intrinsic properties of its *relata*.
18 This solution is partly inspired by Tugby 2013.

also essentially individuated by Cell-2. Anyway, if one does not think of haecceities in terms of modes, (Ψ1) turns out to be more acceptable than (Ψ3).

(Ψ2) has two major problems. Firstly, it does not respect the intuition that there are metaphysically possible worlds in which Cell-2 has different properties, since it claims that Cell-2 is uniquely individuated by a certain, really complex conjunctive property P_C. In this respect, you should accept the questionable assumption that every property within P_C is essential to Cell-2, so that conjunctive properties that are slightly different from P_C in other possible worlds individuate objects that are different from Cell-2. In other terms, Cell-2 does not exist in those worlds in which P_C is not instantiated, even if a conjunctive property which is slightly different from P_C is instantiated there. Secondly, there is a problem with P_C that is analogous to the problem presented in section 3: either each conjunction of properties gives rise to a conjunctive property such as P_C (and perhaps to a power), or only certain conjunctions of properties give rise to conjunctive properties such as P_C. Both alternatives are problematic, as we have already noticed.[19] (Ψ5) shares these problems, as long as this solution only specifies the Platonic nature of the properties involved. Anyway, (Ψ5) has the advantage of admitting U_C's Platonic existence even before its instantiation by Cell-2, so that U_C can contribute to the generative power's individuation even before that power's activation.

Finally, (Ψ4) denies the intuition that there are metaphysically possible worlds in which Cell-2 is constituted by different tropes. Moreover, it also denies that those tropes that participate in grounding Cell-2's individuation in the actual world can "live different lives" in other metaphysically possible worlds.

Alexander Bird (2007, 112) claims that unrealized possibilities (i.e., mere possibilia) exist, even if they are only contingently abstract. "Existence" is here substituted by "concreteness" and he seemingly accepts that, while every object exists, not every object is concrete. The disagreement with Meinongians could be merely terminological. Anyway, concreteness implies a certain characterization of existence in terms of having a spatio-temporal location. Meinongians are not necessarily committed to the acceptance of *this* notion of existence: they could accept other, non-equivalent notions. Moreover, Meinongians typically claim that there is no universal feature of existing – or of having being – that could be legitimately attributed to every entity. Their doctrine is *not* expressed by the thesis that there are entities that have being, even if they do not exist. According to Meinongians, there are entities that do not exist – *full stop*.[20]

---

[19] In the latter case, what turns out to ground the existence and/or the fundamentality of a certain conjunctive property such as P_C is a non-existent object.
[20] Of course, "entities" do not only refer to existents.

The two remaining solutions are more radical. According to the first – the "nothing new under the Sun" solution – there is no new object such as Cell-2, regardless of its existence or non-existence. Seemingly "new" objects are only rearrangements of existing, more basic objects – such as sub-atomic particles. I am not inclined to eliminate cells from my ontological inventory. Anyway, in order for this solution to be a coherent alternative to Meinongianism, it should not only deny that cells start to exist as "new" objects, but also that "new" objects at the fundamental micro-physical level of the universe start to exist. In fact, if there were generative powers at that level, they would be essentially individuated by non-existent micro-physical objects.

Finally, non-Meinongians could assert that there are no objects in the universe (or that objects are not fundamental entities), that the only existent entities (or the only fundamental entities) are powers themselves. However, what would powers be within this perspective? How would they provide a satisfactory reduction of generative powers or, more precisely, of seemingly true ascriptions of generative powers? If powers were something like properties (universal or particular), then such a solution would have the same troubles that characterize (Ψ1)-(Ψ5).

5 Some miscellaneous concerns about Meinongianism

In this final section I shall briefly deal with four major concerns about Meinongianism. Following Quine 1948, it can be held that non-existent objects do not have definite identity conditions, that they are somehow indeterminate, so that they should not be accepted within our ontological inventory. Indeterminate objects are not objects at all. Thus, Cell-2 is not a non-existent object, because it is indeterminate, i.e., because it is not an object at all. The generative power to produce Cell-2 cannot be essentially individuated (among other) by Cell-2, since Cell-2 does not have definite identity conditions and it cannot "help" that power with its individuation.

Yet, it is perhaps the case that the indeterminacy of Cell-2 is only epistemic, that we cannot define what is for something to be Cell-2, rather than Cell-3, even if there is a fact of the matter about their distinction even before their coming to existence. In addition, given Cell-1's features, given the identity conditions for cells and given the ways in which cells generate other cells, it is legitimate to claim that Cell-1 can only generate a certain number of cells with certain features. Perhaps we do not know their exact number and we do not know all their features. Yet, this does not affect their "real possibility", i.e., their being a definite number of distinct, non-existent-yet-possibly-existent objects. Moreover, both Cell-2, Cell-3 and all the cells that can be generated by Cell-1 can still be numerically distinct from one each other. Thus, in my perspective, Cell-1 does not simply have the generative power to produce a cell, but it has the generative powers to produce Cell-2, Cell-3, and so on.

Does this imply an overabundance of powers and/or of objects at the fundamental level of the universe? Put in these terms, the question turns out to be rhetorical. There cannot be "overabundance" of entities at the fundamental level of the universe: that level comprehends all and only the entities that it has to comprehend, regardless of our "economical" evaluation. Thus, if generative powers and non-existent objects such as Cell-2 turn out to be irreducible to other entities, they have the "right" to be part of that level, regardless of their being "too many". Furthermore, attempts to analyse seemingly true ascriptions of generative powers turn out to commit non-Meinongians to an indefinite number of other entities: conjunctive properties, aggregates of tropes, haecceities, and so on. Thus, why should we accept an indefinite number of such entities and not accept an indefinite number of non-existent objects?

Two further concerns remain. Firstly, it seems that what exists cannot be grounded, for its existence and/or for its features, on what does not exist. What exists is somehow more fundamental than what does not exist. Thus, Cell-2 cannot ground the nature of an instantiated generative power and it cannot ground one of the features of Cell-2 (its having that power). However, Meinongians could still preserve the primacy of what exists by claiming that all and only existents have irreducible causal powers, so that all and only existents can explain, by their causings, what happens in the universe.

Finally, you cannot get rid of non-existent objects by simply claiming that such objects would both exist (since there *are*, i.e., there *exist* such objects) and do not exist. The argument that concludes from Meinongianism to this paradox of non-existence is question-begging, since it assumes that *"there is* an object" and *"there exists* an object" have the same meaning – and this is precisely what Meinongians deny!

I shall conclude this paper with one final suggestion. Tugby 2013 claims that, if we accept that there are irreducible powers towards something, we should also accept a Platonic conception of properties. In fact, we should accept that there are properties that are not instantiated – or properties that are still not instantiated. Yet, within the Meinongian perspective that I have defended here, Armstrong's Aristotelianism might be vindicated: seemingly non-instantiated properties could turn out to be properties that are actually instantiated (perhaps in certain peculiar ways) by objects that do not exist.

Bibliography

Armstrong, D. M. 1978. *Universals and Scientific Realism. Volume II: A Theory of Universals.* Cambridge: Cambridge University Press.
Armstrong, D. M. 1982. *What is a Law of Nature?* Cambridge: Cambridge University Press.
Armstrong, D. M. 1995. "Reacting to Meinong". *Grazer Philosophische Studien*, 50, 615–627
Armstrong, D. M. 1997. *A World of States of Affairs.* Cambridge: Cambridge University Press
Armstrong, D. M. 2004. *Truth and Truthmakers.* Cambridge: Cambridge University Press
Armstrong, D. M. 2005. "Four Disputes about Properties". *Synthese*, 144, 3, 309–320
Armstrong, D. M., Martin, C. B., Place, Ullin T. (2002). *Dispositions. A Debate.* New York: Taylor & Francis.
Bird, A. (2007). *Nature's Metaphysics. Laws and Properties.* Oxford: Oxford University Press.
Cartwright, N. 1999. *The Dapple World. A Study of the Boundaries of Science.* Cambridge: Cambridge University Press.
Choi, S., Fara, M. 2012. "Dispositions". In *Stanford Encyclopedia of Philosophy Online*, URL = <http://plato.stanford.edu/entries/dispositions>.
Ellis, B. 2001. *Scientific Essentialism.* Cambridge: Cambridge University Press.
Lowe, E. J. 2006. *The Four-Category Ontology. A Metaphysical Foundation for Natural Science.* Oxford: Oxford University Press.
Martin, C. B. 2008. *The Mind in Nature.* Oxford: Oxford University Press.
Maurin, A.-S. 2013. "Tropes". In *Stanford Encyclopedia of Philosophy Online*, URL = <http://plato.stanford.edu/entries/tropes>.
Molnar, G. 2003. *Powers. A Study in Metaphysics.* Oxford: Oxford University Press.
Mumford, S. 1998. *Dispositions.* Oxford: Oxford University Press.
Mumford, S. 2004. *Laws in Nature.* New York: Routledge.
Mumford, S., Anjum, R. L. 2011. *Getting Causes from Powers.* Oxford: Oxford University Press.
Quine, W. V. O. 1948. "On What There Is". *Review of Metaphysics*, 2, 5, 21–36.
Tugby, M. 2013. "Platonic Dispositionalism". *Mind*, 122, 486, 451–480.

Tuomas E. Tahko
Armstrong on Truthmaking and Realism

1 Introduction

The title of this paper reflects the fact truthmaking is quite frequently considered to be *expressive of realism*. What this means, exactly, will become clearer in the course of our discussion, but since we are interested in Armstrong's work on truthmaking in particular, it is natural to start from a brief discussion of how truthmaking and realism appear to be associated in his work. Armstrong's interest in truthmaking and the integration of the truthmaker principle to his overall system happened only later in his career, especially in his 1997 book *A World of States of Affairs* and of course the 2004 *Truth and Truthmakers*. Since the 2004 book is the most complete account of Armstrong's thinking with regard to truthmaking, that book will be our primary source (especially given that he changed his mind about a few issues between the 1997 and 2004 books). The theme is certainly present in his earlier work as well, but the notion of truthmaking, which Armstrong got from C. B. Martin, was not as well formulated in the literature. The seminal paper by Mulligan, Simons and Smith (1984) had not yet popularised the notion. In the introduction to his *Truth and Truthmakers*, Armstrong outlines the origins of the notion in Australia: Martin used the idea of truthmaking in his work against counterfactual accounts of material objects due to the phenomenalists. Armstrong himself first took advantage of the truthmaker principle in his attempt to resist dispositional/subjunctive accounts of mental states due to behaviourists such as Gilbert Ryle (Armstrong 1973, 11ff.). The now famous slogan, according to which the truthmaker insight 'prevents the metaphysician from letting dispositions "hang on air"' originates in Armstrong's criticism of Ryle (Armstrong 2004, 3).

It is thus partly because of this historical usage of the 'truthmaker insight' that truthmaking is often associated with realism. But the situation is certainly more complicated than that when we look into the details. One reason for the complications regarding truthmaking and its potential ability to capture realist intuitions is that many of the best known theories of truthmaking are very closely tied to certain ontological views that already make realist commitments. Among these is Armstrong's own version of truthmaking, which is integrated with his ontology of states of affairs. Naturally, the states of affairs ontology has some important implications for his conception of truthmaking. The most obvious of these implications

---

Tuomas E. Tahko: University of Helsinki, email: tuomas.tahko@helsinki.fi

is that, according to Armstrong, truthmakers are facts – albeit he prefers to call them states of affairs: 'entities having such forms as a's being F and a's having R to b' (Armstrong 2004, 18). Of course, as Armstrong (2004, 4) readily admits, the idea of truthmaking can be separated from the question of what truthmakers in fact are. In any case, for Armstrong the truthbearers are true propositions – although there are some caveats, e.g., he considers propositions to be 'possible intentional objects' and takes it that a 'naturalist' cannot accept a realm of propositions (*ibid.*, 16; 1997, 131).

Moving on to the truthmaking relation, there are a couple of things that, I believe, can be said without much controversy. One of these is that whatever we take the actual truthmakers to be, and, I suppose, even regardless of the nature of the supposed truthmaking relation between propositions and reality, we can in any case say that the (possible) correspondence between a proposition and the reality, i.e., between the proposition and the truthmaker, is not, in general, a one-one correspondence.[1] This is the view that Armstrong (2004, 16) takes and, in essence, seems to be what many other proponents of truthmaking would go for as well (see for example Lowe 2006, 182). The reason for opting for a many-many relation is simple enough: a single truthmaker can quite clearly be a truthmaker for several truthbearers and correspondingly there might be several truthmakers which serve as a sufficient truthmaker for a given proposition. Perhaps it could be argued that there is always some *minimal* truthmaker for each truth, but as Armstrong points out, many truths do also have several minimal truthmakers, such as the proposition <there exists an x such that x is a human being> (Armstrong 2004, 21).[2] Another aspect that appears to be fairly uncontroversial is that truthmaking is some kind of an asymmetrical relation between propositions and something in the world. This something in the world could be facts or states of affairs, as in Armstrong's case, or tropes, or something quite different, depending on your account of truthmakers. Another way to put this is to say that a truthmaker for a particular truth is some portion of reality in virtue of which the truth is true. This 'in virtue of' relation is generally thought to be cross-categorical, the portion of reality being some entity or entities and the other being truth (which is not an entity!) (see Armstrong 2004, 5).

The exact nature of the truthmaking relation is not uncontroversial though: one possibility is that it is an *entailment relation* between the truthmaker and the truth of the proposition, but it has also been argued that we are dealing with a

---

[1] Note that 'propositions' could be considered merely as a placeholder here, depending on one's take on what the truthbearers are.
[2] Where the angled brackets describe a proposition, following Horwich 1998. For discussion on minimal truthmakers, see Tahko and O'Conaill forthcoming.

grounding relation here, in which case truth would be grounded in entities (cf. Rodriguez-Pereyra 2005). There is also the question of whether truthmaking is an *internal* or an *external* relation (Armstrong 1997, 115-116). Armstrong favours the first alternative, and it does perhaps seem initially more plausible that truthmaking is an internal relation, but there are various problems with this idea as well (cf. David 2005). Each of these issues would require a paper of its own, but we will mostly set them aside here, focusing instead on the more general question regarding truthmaking and realism. However, we will need at least an initial formulation of the truthmaking relation to get started. Take one formulated in terms of the 'in virtue of' locution, which produces a familiar truthmaker principle:

(TM) Necessarily, if a proposition <p> is true and has a truthmaker, then there is some entity in virtue of which it is true.

This formulation of (TM) entails (though is not the same as) *Truthmaker Necessitarianism*: the existence of a truthmaker is sufficient for the truth of those propositions it makes true. Armstrong (2004, 5-7) defends Truthmaker Necessitarianism, appealing to the slogan mentioned above, i.e., if a given truth (a true proposition) would lack a truthmaker, then its truth would 'hang on air' quite like Ryle's dispositional truths. Indeed Truthmaker Necessitarianism is a widely shared assumption amongst truthmaker theorists, even though it is difficult to come up with a conclusive argument in favour of it:

> I do not have any direct argument [for Truthmaker Necessitarianism]. My hope is that philosophers of realist inclinations will be immediately attracted to the idea that a truth, any truth, should depend for its truth for something 'outside' it, in virtue of which it is true. (Armstrong 2004, 7.)

We will not discuss Truthmaker Necessitarianism in much more detail than this, nor the other one of Armstrong's controversial theses, *Truthmaker Maximalism*, i.e., the thesis that *every* truth must have a truthmaker.[3] (TM) does not, of course, entail Truthmaker Maximalism.

2 Truthmaking and realism

It is well-known that Armstrong postulates an intimate connection between truthmaking and realism. The mediator here is *correspondence*, or more precisely, the

---

[3] For a brief defence of Truthmaker Maximalism, see Rodriguez-Pereyra 2006.

correspondence theory of truth. It appears that, for Armstrong, truthmaker theory could be understood simply as a more sophisticated version of the correspondence theory. There are several passages in Armstrong's work that explicitly suggest this:

> [T]he Correspondence theory tells us that, since truths require a truthmaker, there is something in the world that corresponds to a true proposition. The correspondent and the truthmaker are the same thing. (1997, 128)

> *Propositions correspond or fail to correspond to reality.* [If Armstrong's view of propositions is correct], then it becomes pretty clear that the correspondence theory of truth can and should be upheld. (2004, 16)

> The terms of the correspondence relation are truthmakers and truths. Truthmakers entail truths. Our favoured truthmakers are states of affairs or their constituents. (1997, 131)

Note however that especially in *A World of States of Affairs*, Armstrong emphasises that truthmaker theory is not only compatible with (the idea of) the correspondence theory of truth, but also with the redundancy theory of truth (if it is to be called a theory of truth at all) (1997, 128).[4] Be that as it may, in the secondary literature Armstrong's appeal to the correspondence theory in his formulation of the truthmaker theory has been received with some hostility. Typically, this is because the motivation seems to be exactly to argue in favour of realism. Consider how Helen Beebee and Julian Dodd put it in their influential volume on truthmaking:

> Suppose that some formulation of truthmaker theory does indeed succeed in capturing realist intuitions. The question arises, how can truthmaker theory now legitimately be put to use in an argument *for* realism (about a particular domain) and *against* anti-realism? If truthmaker theory itself enshrines a commitment to realism, then presumably the appropriate anti-realist reaction to such an argument is simply to deny whatever truthmaker principle is being used as a premise in that argument. If a given truthmaker principle is to pull its weight in arguments against anti-realism, then we had better have reasons, independently of our commitment to realism, for believing that the principle is true. We wonder whether such reasons are to be had. (Beebee and Dodd 2005, 16.)

---

4 Incidentally, one might also ask whether the truthmaker theory is a *theory* of truth. Certainly, if we have merely a stripped down truthmaker principle, then we are not dealing with a complete theory of truth. But once the truthmaker principle is combined with an appropriate ontology, then I would be inclined to say that we do indeed have a complete theory of truth, as we can give a full account of the truthbearers and the truthmakers. But the core of truthmaking is the truthmaker principle, and if it turns out to be compatible with different ontologies (as will be proposed below), then it is at least a promising starting point for a complete theory of truth.

So if Armstrong's postulated connection between truthmaker theory and realism truly holds, then Beebee and Dodd insist that we should be able to put forward a truthmaker principle that would be able to capture our realist intuitions while not being compatible with anti-realism. Even if we were to succeed, we would still have to show that there are reasons, independently of our realist intuitions, to believe that our truthmaker principle is the correct one, as otherwise the use of truthmaking in arguments against anti-realism will just be question-begging. Perhaps this can be done, but as we have seen, Armstrong's own project seems to postulate a very intimate connection between truthmaking and realism.

However, if we were to concede that truthmaker theory fails to cash out our realist intuitions, at least without leaving room for *other* interpretations, then what would the cost be, precisely? Well, provided that truthmaker theory is at least *compatible* with realism – which it surely is – then the possibility that it might be able to accommodate other than realist intuitions might not be so harmful. In other words, if truthmaking turned out to be an *ontologically neutral* way of talking about truth we could of course still combine it with a realist ontology. Now, this is of course not an answer to the challenge posed by Beebee and Dodd. Rather, the proposal is that the price that Armstrong might have to pay is not all that high. But this line of argument is only feasible once it is clear that truthmaker theory *can* be presented in an ontologically neutral way and if realism itself can stand on its own. So let us now move to a discussion of truthmaker theory combined with various alternatives to realism.

3 Truthmaking as ontologically neutral

If we wish to find an ontologically neutral formulation of the truthmaker principle, then this is likely to impose some constraints on the theory. For that reason, certain usual formulations are unlikely to work. Consider one typical formulation, as presented by Beebee and Dodd:

(TM-E) Necessarily, if <p> is true, then there exists at least one entity a such that <a exists> entails <<p> is true>. (Beebee and Dodd 2005, 2.)

The nature of the truthmaking relation, here suggested to be an entailment relation, is perhaps the most controversial part of (TM-E). Of course, other problems may emerge when certain truths, such as necessary truths or negative truths are considered. There have been numerous attempts to deal with these problems, but the details of each solution depend, often heavily, on the details of the ontology that one wishes to combine with truthmaking, and accordingly these problems are not something that we should focus on here. However, a somewhat neutral way to address the problems involved with entailment is to replace entailment with (metaphysical) necessitation: in every possible world where a truthmaker for a certain proposition exists, that proposition is true. This is the line that was taken in the initial formulation given in this paper (TM) and it would seem to be preferable to Armstrong (1997, 115) as well (see also Lowe 2006, 185).

Some key features of the truthmaker principle were listed earlier and at least some of them would also seem to hold in regard to the general principle that we are now looking for. So, we can for example without much risk of controversy say that truthmaking is an asymmetrical many-many relation. Also, as Rodriguez-Pereyra (2005, 20-1) suggests, we seem to have the intuition that truth is asymmetrical, and the truthmaker principle fits this intuition perfectly. The way that Rodriguez-Pereyra puts it is that truth is grounded: the truth of a proposition depends on what reality is like, and the relationship between truth and reality is of course asymmetrical, for reality does not depend on the truth of the proposition. As he points out, this by itself does not commit us to realism, for an idealist (for instance) could just add that reality or world and the entities in it are not mind-independent (*ibid.*).

Moreover, Chris Daly (2005) has suggested that there is one issue that advocates of different truthmaker theories always agree upon: truthmaking does some *explanatory* work. This is of course a rather natural source for motivation to adopt truthmaker theory in the first place. Ultimately, this motivation concerns the nature of the truthmaking relation, for whatever explanatory work the truthmaker principle might do, it must surely have something to do with the relationship between propositions and truthmakers. So what are our options for motivating truthmaking? According to Daly (2005, 102), there are three options. The first one is what he calls the 'Canadian mountie' theory of truthmakers, the idea of which is to argue from examples and to show that we can, in fact, *always* find a truthmaker for any given truth. Daly accuses this theory of being *ad hoc*, in that it assumes the truthmaker principle without giving any justification for it. Presumably the point is that we need more than a working theory of truthmaking to motivate the idea in the first place. This would appear to be a valid request.

The second strategy suggests that truthmaker theory could help in finding explanations to further ontological problems, such as the theory of universals. Daly (2005, 98-102) argues against Rodriguez-Pereyra's suggestion, namely that truthmakers could explain universals by entailing that it is true that there are some properties which are shared by several distinct particulars. There are other alternatives as well though, one of them being Josh Parsons's (2005) rather plausible idea that truthmaking could be used to motivate arguments concerning propositions about the past and the future and thus might provide some explanatory power when discussing theories of time, such as presentism. However, while I am not against the idea of granting the possibility that truthmaking could help motivate arguments concerning other ontological problems, I do not believe that this by itself is a sufficient condition for adopting the truthmaker principle; and neither, of course, does Daly.

The third strategy that Daly (2005, 94–8) considers, namely inference to the best explanation, is perhaps the most common. According to this strategy, truthmaking explains our pro-realism intuitions and grasps the core idea of the correspondence theory of truth. This is of course the core motivation that we are now interested in. Daly considers Armstrong's and Bigelow's theories in this connection. Here we are faced with the central question: could truthmaking offer a way to characterise a theory of truth compatible with realism? But we have to be careful here, for even if truthmaking is compatible with realism, it does not mean that it would explain why realism is any better than other alternatives. Indeed, it seems that the truthmaker principle is in no way connected with any necessarily realist premises, especially if it is compatible with, say, pragmatism and idealism as well (which is a suggestion we will consider briefly below).

There is one further complication. Recall that Armstrong seems to consider truthmaking to be effectively a more sophisticated version of the correspondence theory – and this is in fact a major reason for the claim that it captures realist intuitions. If this were indeed the case, then it would seem difficult to combine truthmaking with anything but realism. But this is where things get interesting, for Daly argues that the same *ontological neutrality thesis* applies to the Correspondence Intuition (CI) as well, formulated in the following way:

(CI)   <p> is true if and only if things are as <p> says they are. (Daly 2005: 96.)

The apparent problem with (CI), however, is that it appears to be vacuous: (CI) is compatible with just about any theory of truth, and hence its explanatory value cannot be particularly high. So, if truthmaker theory is supposed to be explanatory, it better capture something more than just (CI). Armstrong himself (2004: Ch. 4) certainly claims that the truthmaker principle could say something more than (CI) does – this will be done by combining the correspondence relation with the truthmaker principle and his states of affairs ontology – but consider Daly's analysis of (CI):

> Consider the coherence theorist. He may consistently say 'If <p> is true, it has a truthmaker. <p> corresponds to a state of affairs, namely the state of affairs which consists of a relation of coherence holding between <p> and the other members of a maximal set of propositions'. Consider the pragmatist. He may consistently say, 'If <p> is true, it has a truthmaker. <p> corresponds to a state of affairs, namely the state of affairs of <p>'s having the property of

being useful to believe'. It is controversial whether there exist states of affairs. Let that pass. My point here is that the coherence theory and the pragmatic theory are each compatible with the admission of states of affairs. Furthermore, each of these theories is compatible with the admission of states of affairs standing in a correspondence relation to truths. (Daly 2005, 97)

So Daly's case against the third strategy (to guarantee the explanatory value of truthmaker theory and hence motivate it) is based on the claim that the truthmaker principle does not restrict our choices in terms of ontology in any way and thus truthmaking understood in the lines of Armstrong and Bigelow is just as vacuous as (CI). This is indeed a valid concern, for if truthmaking is understood as a special case of the correspondence theory, then it seems to inherit all of its original problems.

However, it seems trivial that the truthmaker principle could be combined with different ontologies once we acknowledge the idea that truthmaking is quite separate from the varying answers concerning the actual truthmakers and truthbearers. Furthermore, as already noted, Armstrong (2004: 4) seems to have no quarrel with the idea that truthmaking may be compatible with very different accounts of truthmakers and truthbearers. This is really the only thing that counts: it ought to be one's account of truthmakers and truthbearers that introduces the (important) ontological commitments, not the truthmaker principle itself. Accordingly, I think that Armstrong and other advocates of realist truthmaker theories could very well be content with a somewhat weakened condition when it comes to the truthmaker principle, namely, that the truthmaker principle is the best way to characterise the correspondence relation understood *in a realist sense*. When put like this, the details of our ontology are still open, but the motivation for truthmaking is still clear: it is the best way to formulate the realist understanding of the correspondence relation. This hints towards a fourth strategy for motivating truthmaking in addition to the three suggested by Daly, and in fact I think that the fourth strategy is closer to how most truthmaker theorists would like to motivate their theories.

4 Realism can stand on its own

The strategy for motivating truthmaking that is now emerging rests on this very simple point: realism can stand on its own. In other words, we do not need truthmaking (or the correspondence theory, for that matter) to motivate realism. This reflects Michael Devitt's (1997) classic work on the topic of realism and truth. Compared to Daly's third strategy, this changes the direction of explanation. It could be said that the fourth strategy does not so much try to provide an explanation, but a justification, although in another sense it can be thought to provide an explanation as well, as we will shortly see. In any case, what is important is that because realism can stand on its own, those of us who are realists can motivate truthmaking with realism – not the other way around. While this type of strategy is not clearly present in Armstrong's writings, I do believe that he might have welcomed it.

If we start with a realist ontology and if truthmaking increases the plausibility of the overall theory, then it seems rather straightforward to choose the way to go: realism plus truthmaking is the best theory available. But in order for this strategy to be plausible, we ought to see some more evidence to the effect that, say, idealists or pragmatists would also be happy with the proposed truthmaker principle. To this effect, the principle would have to be such that an idealist or pragmatist could insert their desired truthmakers and truthbearers into the principle. There may be some limitations here. For one thing, on all the usual formulations, the truthmakers are taken to be entities of some kind. It is certainly a matter of debate what kind of entities they are, but it might be objected that, say, a pragmatist would not be happy about the commitment to 'entities' in any form whatsoever. So how could pragmatism be compatible with truthmaking? Positive accounts arguing to this effect are scarce, but Sami Pihlström (2005) has outlined some options. Pihlström suggests that pragmatists such as Hilary Putnam (at times) and Nelson Goodman could very well be considered as taking advantage of a version of truthmaking, whereby the truthbearers and the 'world' that makes them true (i.e., the truthmakers) are human constructions, 'made' by us in the process of representing and acting. Now, whether this constitutes a commitment to entities or not is perhaps debatable – maybe 'human constructions' are to be considered as entities. But there appears to be no reason why this type of picture couldn't be represented with a truthmaker principle not unlike the ones we have been discussing.

A more general point to note is the following. If we take truthmakers to be entities, there are several alternatives available, such as Armstrong's states of affairs or tropes, as suggested in Mulligan, Simons and Smith (1984). There is not much that can be said about the nature of the truthmakers without a commitment to a particular ontology. However, personally I would be inclined to part ways with Armstrong here, for it seems to me that the apparent complexity of truth suggests that truthmakers must be spread out in several different categories rather than just one. This complexity manifests itself in the variety of things we consider to be true: mathematical theorems, laws of physics, that Hesperus and Phosphorus are identical, and so on. Introducing a further category of facts or states of affairs to account for all truthmakers is not ontologically parsimonious; why not say that the truthmakers are just the very entities that a given true proposition concerns? This line of thought has also been noted by Beebee and Dodd (2005, 9) and it is exactly what Lowe (2005, 182 ff.) argues for as well.

Returning to the issue concerning a commitment to entities, we might note that one could attempt to formulate truthmaking in a manner that does not entail a commitment to entities at all, perhaps in the lines of McFetridge's (1990) suggestion that every true sentence must have an explanation of *why* it is true. This would seem to release us from the commitment to entities, but it also distances us from the original idea of truthmaking. In fact, it appears that this would take us back towards a vacuous principle. McFetridge's proposal is another attempt to combine truthmaking and our realist intuitions so that we would have an argument against anti-realism (see Liggins 2005 for details of how this might be done). This is a line of thought that we have already distanced ourselves from. Of course, we can easily modify the truthmaker principle in such a way that an explicit commitment to entities is removed, but it is questionable whether this really does the trick. Consider (TM*):

(TM*) Necessarily, if a proposition <p> is true and has a truthmaker, then there is some a in virtue of which it is true.

Here the entailment relation has been replaced with metaphysical necessitation, as in the original (TM). In fact, you'll notice that (TM*) differs from the original (TM) only in replacing the explicit reference to entities with a reference to the unspecified 'a'. Have we been moving in circles? Not as such: what has changed is the order of explanation. Truthmaking is now understood as a tool to help characterise one's ontology, not a way to motivate the ontology itself. To that effect, all we need is that the truthmaker principle is compatible across different ontologies. (TM*) is obviously compatible with realist ontologies, in which case it is likely that we would want to add that what makes <p> true is the existence of an entity of some kind. A pragmatist, on the other hand, could replace a with 'human construction', as suggested by Pihlström. Whether or not this avoids the commitment to entities is another question, but one that I consider an ancillary issue. As for idealists, they could presumably interpret existence so that it does not require *material* existence, although I am not aware of any idealist accounts which would employ truthmaking explicitly.

Naturally, we need to add something to (TM*) to give it any explanatory power, as the nature of the truthmaking relation depends on what a is – and now also on how we interpret *existence*. Indeed, (TM*) is just the spine of truthmaking as we need to say something about a to determine *how* it makes <p> true. However, the relative weakness of (TM*) is exactly why the principle is plausible across ontologies: it could perhaps be interpreted as a family of relations that covers all possible kinds of truthmakers. Of course, in (TM*) the truthbearers are still taken to be propositions, which might not satisfy everyone (or every ontology). But I would not be too concerned about this, given that Armstrong himself wishes to avoid a commitment to propositions and reinterprets them as 'possible intentional objects', in accordance with his naturalistic agenda. Accordingly, perhaps there is room to interpret propositions as well in accordance with various ontologies.

The upshot is that truthmaking is not, or does not have to be, an explanation for, or a case in favour of our realist intuitions. Perhaps truthmaking does increase the appeal of realism, for the explanatory power of the complete theory (realism plus truthmaking) is certainly greater with truthmaking than without it. In this sense, truthmaking may still make a contribution towards explanatory power. But an idealist or a pragmatist could attempt to make the same claim. In any case, there are strong reasons to think that the question of realism is independent of the question of truth. Armstrong may have wanted something more than this out of truthmaker theory, nevertheless, truthmaker theory is a plausible way to account for truth within a realist ontology. A particularly forceful reason to think so is that, in the lines of Michael Dummett's (e.g. 1991) influential work, the anti-realist's strongest case against realism may be exactly that realism is unable to account for truth in a satisfactory manner, given the shortcomings of the correspondence theory. If truthmaker theory can now offer an ontologically neutral way to account for truth, then this argument dissipates. Realism can stand on its own, and truthmaking is a way to account for truth regardless of one's ontology. This is really all that Armstrong or any realist proponent of truthmaker theory needs: a way to account for truth within a realist ontology.

Bibliography

Armstrong, D. M. 1973. Belief, Truth and Knowledge. Cambridge: Cambridge University Press.
Armstrong, D. M. 1997. A World of States of Affairs. Cambridge: Cambridge University Press.
Armstrong, D. M. 2004. Truth and Truthmaking. Cambridge: Cambridge University Press.
Beebee, H. and Dodd, J. (eds.) 2005. Truthmakers: The Contemporary Debate. Oxford: Oxford University Press.
Daly, C. 2005. "So Where's the Explanation?" In Beebee, H. and Dodd, J. (eds.) 2005, 85–103.
David, M. 2005. "Armstrong on Truthmaking". In Beebee, H. and Dodd, J. (eds.) 2005, 141–59.
Devitt, M. 1997. Realism and Truth, 2nd edn. Princeton: Princeton University Press.
Dummett, M. 1991. The Logical Basis of Metaphysics. Cambridge, MA: Harvard University Press.
Horwich, P. 1998. Truth, 2nd edn. Oxford: Oxford University Press.

Liggins, D. 2005. "Truthmakers and Explanation". In Beebee, H. and Dodd, J. (eds.) 2005, 105–15.
Lowe, E. J. 2006. The Four-Category Ontology: A Metaphysical Foundation for Natural Science. Oxford: Oxford University Press.
McFetridge, I. 1990. "Truth, Correspondence, Explanation and Knowledge". In his Logical Necessity and Other Essays. London: Aristotelian Society.
Mulligan, K., Simons, P. and Smith, B. 1984. "Truth-Makers". Philosophy and Phenomenological Research 44: 287–321.
Parsons, J. 2005. "Truthmakers, the Past, and Future." In Beebee, H. and Dodd, J., 161–74.
Pihlström, S. 2005. "Truthmaking and Pragmatist Conceptions of Truth and Reality". Minerva – An Internet Journal of Philosophy 9. URL = <http://www.ul.ie/~philos/vol9/Truthmaking.html>.
Rodriguez-Pereyra, G. 2005. "Why Truthmakers". In Beebee, H. and Dodd, J. (eds.), 17–31.
Rodriguez-Pereyra, G. 2006. "Truthmaker Maximalism Defended". Analysis 66 (3): 260–64.
Tahko, T. E. and O'Conaill, D. forthcoming. "Minimal Truthmakers". Pacific Philosophical Quarterly.

D. H. Mellor
From Translations to Truthmakers

My metaphysics owes much to David Armstrong. His work has always influenced mine even when we disagree, as what follows will illustrate by relating my views of truthmaking and the mind to those in *A Materialist Theory of the Mind*.

1 Translations

Quine 1948 argues that 'in debating over what there is, there are still reasons for operating on a semantic plane' (p. 16). And so, up to a point, there are: we can hardly tell if neutrinos or unicorns exist without knowing roughly what we mean by 'neutrino' and 'unicorn'. But there is more to ontology than semantics. For a start, the semantically trivial equivalence principle

(EP) 'P' is true if and only if (iff) P

can hardly settle our ontology. As Dyke 2008 notes, the truism that

> 'Eating people is wrong' is true iff eating people is wrong (p. 5)

can hardly make the truth of 'Eating people is wrong' entail the existence of moral properties: the anti-realism of Ayer 1946, the error theories of Mackie 1977, and the naturalism of Foot 1978, are not so easily refuted. And it must take more than the truism that

> 'The sun is a star' is true iff the sun is a star

to make the truth of 'the sun is a star' refute those who 'don't really believe in astronomy except as a complicated description of part of the course of human and possibly animal sensation' (F. P. Ramsey 1925, 249).

Perhaps we can give semantics more ontological clout by using a suitable translation of 'P' to say when it's true. 'Suitable' must of course exclude trivial translations into other languages, like

> 'The sun is a star' is true iff die Sonne ist ein Stern,

D. H. Mellor: University of Cambridge, email: dhm11@cam.ac.uk which tells us no more about what there is than its all-English equivalent. More promising perhaps are the translations Quine 1948 invokes to avoid unwanted 'ontological commitments', for example to abstract entities like species:

> when we say that some zoological species are cross-fertile we are committing ourselves to recognizing as entities the several species themselves [...] until we devise some way of so paraphrasing this statement that the seeming reference to species [...] was an avoidable way of speaking (p. 13).

This test is not trivial, since it may fail: possible paraphrases may all share the original's unwanted commitments or lack wanted ones. Nominalists, for example, cannot paraphrase 'Red is a colour' as 'Necessarily, all red things are coloured', since (a) this doesn't imply that red things resemble each other in colour and (b) all red things are also necessarily extended and shaped (Jackson 1977, 89–90). But even if 'Red is a colour' had a credible nominalistic paraphrase, Quine's test also needs an independent 'argument for preferring the ontological commitments of [the] paraphrase to those of the original sentence' (Dyke, 86) – an argument the paraphrase's mere existence does not provide. Failing such an argument,

> 'Red is a colour' is true iff, necessarily, all red things are coloured

would, even if true, be no help to nominalists.

2 Truth conditions

We can put the point just made in terms of Tarskian truth conditions. Tarski 1944 uses these to protect 'object languages' from the Liar and other paradoxes by deporting their semantic predicates, like 'true' and 'false', into meta-languages that we can then safely use to say when sentences of their object-languages are true. But doing this does not require a meta-language's *non*-semantic predicates to differ from those of its object language. Yet unless some do, a meta-linguistic statement of a sentence's truth conditions, like

> 'Red is a colour' is true iff red is a colour

merely instantiates the trivial equivalence principle (EP). For a sentence's truth condition to be ontologically informative, it must at least use a non-trivial paraphrase, as

> 'Red is a colour' is true iff, necessarily, all red things are coloured,

would do if it was true, that also has an ontological authority its object-language equivalent lacks. That is why, even then, it would take a different, non-semantic, argument for nominalism to justify replacing 'Red is a colour' with 'Necessarily, all red things are coloured' as a less ontologically committing surrogate.

In any case, realism and nominalism, as doctrines that apply equally to all natural properties, tell us nothing specifically about colours. But some truth conditions for 'x is red', 'x is green', etc., do, because part of what we mean by colour terms is that they apply to a way things look to us. So non-colour-blind people can learn what they mean by learning to recognise the looks of things that make us call them – and their looks – 'red', 'green', etc. That is a semantic fact we can then express by saying, of any x made visible by the light it reflects, that

> 'x is red' is true iff x looks red in daylight to non-colour-blind people,
>
> 'x is green' is true iff x looks green in daylight to non-colour-blind people,
> etc.

This may not tell us much about the features of things, light, and our eyes, that makes some things (look to us) red and others green. But it does tell us something, by setting a condition which answers to that question – e.g. about how surfaces reflect light of various visible frequencies – must meet.

Simpler, and more applicable here, is the fact, exploited by pressure cookers, that the boiling point of water, as of other liquids, increases with its pressure, as shown in Figure 1:

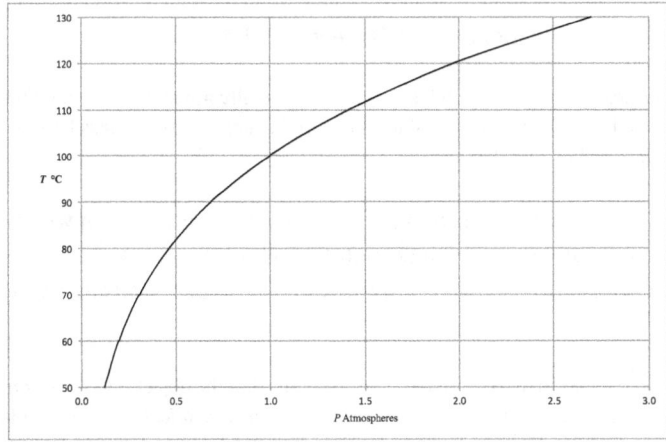

Fig. 1. Boiling points

Thus, if T_P is water's boiling point under pressure P, and P_T is the pressure under which water boils at temperature T, water will boil iff its temperature T and pressure P are such that T is no less than T_P or, equivalently, that P is no greater than P_T. This gives us the following two truth conditions for any sample x of water:

'x boils at T' is true iff x's pressure $P \leq P_T$, and

'x boils at P' is true iff x's temperature $T \geq T_P$,

If these too hardly give us the full facts about water and other liquids that fix their boiling points – e.g. their molecular structures – they do set a condition which those facts must meet. That itself is a non-trivial fact about liquids.

3 Dispositions

Restating our truth conditions for 'x boils at T' and 'x boils at P' in terms of dispositions as

x is disposed to boil at T iff its pressure P is no greater than P_T, and

x is disposed to boil under pressure P iff its temperature T is no less than T_P

may make their ontological implications clearer to us, if not to Ryle 1949, for whom

> dispositional statements are neither reports of observed or observable states of affairs nor yet reports of unobserved or unobservable states of affairs (p. 125).

However, the incredible implications of Ryle's view, e.g. that

> two seemingly identical glasses of which just one is fragile really are identical while they are not being dropped. No event, with possible causes and effects, makes a glass become, or cease to be, fragile (Mellor 1974, 108),

has made me, like Armstrong, a realist about dispositions (1974; 2000) for whom, therefore, dispositional statements can have ontological implications.

My realism about dispositions does not, however, extend to Armstrong's view that

> to speak of an object's having a dispositional property entails that the object is in some *non-dispositional* state or that it has some property (there exists a *categorical basis*) which is responsible for the object manifesting certain behaviour in certain circumstances, manifestations whose nature makes the dispositional property the dispositional property that it is (Armstrong 1993, 86; *my italics*).

Armstrong's dispositional/categorical dichotomy has been rejected by others too, on various grounds (Armstrong et al. 1996) not all of which I accept. My objections to it are in my 1974, §5, and 2000, §4, where I argue that dispositions need *no* non-dispositional bases to be 'categorical', i.e. real, properties. Take for example inertial mass, as defined by Newton's laws of motion (Newton 1713 vol. 1, 13). These laws entail that the inertial mass Mx of any solid object x, any net force Fx applied to it, and x's consequent acceleration Ax, will satisfy

$$Mx = Fx/Ax$$

(provided M, F and A are measured in consistent units, e.g. kilograms, newtons, and metres/second^2). This makes Mx a many-valued function of Fx and Ax, just as we saw in §2 that water's boiling or not is a two-valued function of T and P. And just as that function entails the two dispositions given in §2, so Newton's function entails that

> x is disposed to accelerate at A when any net force F is applied to it iff its mass M is F/A, and
>
> x is disposed to accelerate at A when its mass is M iff the net force F applied to it is A/M.

But if we now ask, as in §2, what property or properties of a solid object x gives it all these dispositions, the only possible answer is: its inertial mass. That dispositionally-specified property just *is* the property 'responsible for the object manifesting certain behaviour in certain circumstances', in this case the rate and direction in which any net force will accelerate it.

4 Truthmakers

That a *single* property is responsible for all these independent 'manifestations' of inertial mass is an even more striking ontological fact about solids than the existence of boiling points is about liquids. It explains the myriad and otherwise inexplicable coincidences entailed by the fact that

$$Mx = Fx/Ax$$

quantifies over infinitely many different directions and forces. It explains them by making 'x has mass Mx' entail every conditional of the form

F accelerates x at A iff nF accelerates x at nA,

in every direction, and for every F, x, and real n for which nF doesn't alter Mx (e.g. by knocking bits off x.)[1] An object x's inertial mass Mx is what makes all these logically independent and infinitely numerous conditionals true (or, if you prefer, truth-preserving): it is their *truthmaker*.

Carnap 1936–7 gave a different counter-example to Armstrong's dispositional/categorical dichotomy decades before Armstrong asserted it:

> The intensity of an electric current can be measured [...] by measuring the heat produced in the conductor, or the deviation of a magnetic needle, or the quantity of silver separated out of a solution, or the quantity of hydrogen separated out of water etc. We may state a set of bilateral reduction sentences,[2] one corresponding to each of these methods. The factual content of this set is not null because it comprehends such sentences as e.g. 'If [a] the deviation of a magnetic needle is such and such then [b] the quantity of silver separated in one minute is such and such, and *vice versa*' which do not contain the term 'intensity of electric current', and which obviously are synthetic (p. 56).

Each constituent – e.g. [a] and [b] – of Carnap's synthetic conditionals specifies a natural property, namely the intensity I of an electric current e such that

e is I iff [a] e is disposed to make a magnetic needle deviate by N degrees;

e is I iff [b] e is disposed to separate M grams of silver per minute from a solution.

But it is only because both conditionals specify the *same* I that, as Carnap says, they entail that

if [a] the deviation of a magnetic needle is N then [b] the quantity of silver separated in one minute is M, and *vice versa*.

And similarly for all Carnap's other conditionals: their all specifying the same intensity I is what makes the fact that a current e has that intensity the truthmaker for all of them.

In these as in other cases the more independent dispositional specifications a theory can use to identify a single property, the more conditionals instances of that property will make true, and the more therefore the theory will explain. Thus

---

[1] Those who deny that conditionals have truth values may read 'S entails C', where C is 'If A then B' and S, A and B are not conditionals, as 'Necessarily, C preserves truth if S is true'.

[2] Carnap's reduction sentences are material conditional readings of what I call dispositional specifications.

adding Carnap's '[a] iff [b]' to Maxwell's electromagnetic theory (1873) enables that theory to explain the phenomena of electrolysis by taking the intensity I of a current e to be the truthmaker for

> e is disposed to separate M grams of silver per minute from a solution.

Similarly, the identification of the *inertial* mass of objects, defined as above, with their *gravitational* mass, enabled by

> the existence of a field of force, namely the gravitational field, which possesses the remarkable property of imparting the same acceleration to all bodies (Einstein 1923, 114),

allows the general theory of relativity to explain gravitational phenomena, not by gravitational forces, but by making the masses and spatiotemporal relations of objects the truthmakers for truths about the variable curvature of spacetime, which then determines how objects move under gravity.

Yet the truth of neither of these theories, of electrical and gravitational phenomena, requires the dispositionally-specified property that provides its truthmakers to have a non-dispositional basis. Why then require other dispositions to have them? Why, in particular, require the dispositionally-specified beliefs and desires that cause our actions to do so?

5 Functionalism

It is a truth universally acknowledged that we mostly do what we believe will get us what we want. Suppose for example I go straight to my pub because I come to believe it's open and want a drink. That belief and that desire cause me to go by being what Mackie 1965 calls 'INUS conditions'[3] of my action: that is, causes each of which needs the others to make it necessary and sufficient for its effect. These two causes of my going to my pub are not of course its only INUS conditions: two others are that I am physically able to get to my pub, and that I believe it has the drink I want. But if all my going's other INUS conditions, whatever they are, are met, then whether I go becomes a function of whether I desire a drink and whether I believe that my pub's open, just as whether water boils is a function of its temperature and pressure. And just as that function can be restated as two dispositions, so can this one:

---

[3] Mackie defines an INUS condition as 'an insufficient but necessary part of a condition which is itself unnecessary but sufficient for the result' (p. 34).

I am disposed to go straight to my pub when I want a drink iff I believe my pub's open;

I am disposed to go straight to my pub when I believe it's open iff I want a drink.

These dispositions do not of course suffice to distinguish this belief and desire from all my other beliefs and desires, if only because many of those – e.g. my belief that I have enough money to buy my drink – are also INUS conditions of my action. But they do tell us something about the two states of mind they cite; and the wide range of actions that these mental states would make others cause tells us more. If for example I believe I need more money, the belief that my pub's open will dispose me to go to it not straight but via a cash point; if I don't want anything from my pub, believing it's open will not dispose me to go to it at all; and so on. Many other beliefs and desires will make any one belief or desire give me, or deprive me of, quite different dispositions.

And besides all the behavioural dispositions our beliefs and desires give us, we have their interactions (e.g. believing my pub's open causing me to want a drink), and the perceptual causes of our beliefs (e.g. seeing someone walk into my pub causing me to believe it's open). Between them, the causes, interactions and behavioural effects of our contingent beliefs and desires may even suffice to identify all of them. The thesis that they do is the functionalism which Braddon-Mitchell and Frank Jackson (2007, ch. 3) call 'common sense' or 'analytic' but I, for obvious reasons, prefer to call 'causal'. It is an application to mental states of an extended version of the thesis that

> what makes a property the property it is, what determines its identity, is its potential for contributing to the causal powers of the things that have it (Shoemaker 1980, 234).[4]

It is also a natural extension of Armstrong's thesis that

> the concept of a mental state is primarily the concept of *a state of the person apt for bringing about a certain state of behaviour*. Sacrificing all accuracy for brevity, we can say that, although the mind is not behaviour, it is the *cause* of behaviour. In the case of some mental states only they are also states of the person apt for being brought about by a certain sort of stimulus (1993, 82, *his italics*).

---

[4] This, *pace* Shoemaker (§§7–9), does not entail his further claim, fortunately irrelevant to what follows, that 'all of the causal potentialities of a property are essential to it' (p. 240). See Mellor & Oliver 1997, 30–31.

6 Physicalism

But if causal functionalism is true, and we can distinguish beliefs and desires by their behavioural effects, perceptual causes and mutual interactions, why go on to identify them with 'physico-chemical states of the brain' (Armstrong 1993, 90)? Armstrong's answer is that these brain states are the non-dispositional states that *cause* the behaviour he uses to define mental states: hence his materialism. And despite denying that dispositions need non-dispositional bases, I could still give his answer, if not for his reason. For if properties like mass and the intensity of electric currents can be identified by their 'potential for contributing to the causal powers of the things that have [them]', as we saw in §4 they can, so can 'physico-chemical states of the brain'. And if currents can have the wide range of effects listed in §4, as they do, why can those brain states not have all the 'behavioural effects, perceptual causes and mutual interactions' of our beliefs or desires?

Two well-known facts about our beliefs and desires may further motivate their identification with brain states. If I believe it's daylight because I can see it is, the light that causes my belief must do so by affecting my eyes, optic nerves, and thence my brain; and similarly, *mutatis mutandis*, when our other senses cause beliefs. We know too that the external behavioural effects of our beliefs and desires require bodily intermediaries: *pace* Uri Geller, we can only bend spoons by first moving our muscles. Are not these facts best explained by identifying beliefs with the brain states that perceptions cause, and which in turn cause the bodily mechanisms of our behaviour?

I say not, but not because I am a dualist. My objections to identifying beliefs and desires with brain states are not objections to physicalism, as is shown by their applying equally to the theory that the temperature Tx of any gas sample x is the mean kinetic energy of x's molecules (Kripke 1971). Of the reasons given in my 2000, §§9–10, for rejecting this inexplicably popular theory, the most obviously applicable here is that

> there is in reality no property of *mean* kinetic energy to identify temperature with, any more than there are in reality the 2.4 children that average families have: there are only the actual kinetic energies of individual molecules, whose mean value is what the kinetic theory relates to a gas's temperature (p. 93).

This makes temperature and kinetic energy properties of different things: the former of gas samples, the latter of gas molecules. And even if mean kinetic energy *was* a property, of groups of molecules as well as of single ones, it would still not be co-extensive with temperature. For if it was, then when a single molecule of a gas sample at room temperature happens to be at rest (e.g. while bouncing off another molecule), with zero kinetic energy, it would be at absolute zero, which it isn't; and speeding it up would automatically heat it up, which it won't. Yet unless temperature and mean kinetic energy *are* coextensive, as inertial and gravitational mass are, they cannot be identical, on pain of contradiction: nothing can simultaneously both have and lack a single property.

But then, *pace* Kripke, no version of kinetic theory ever said that temperature and mean kinetic energy were identical, any more than the simple gas law,

$$Tx = kPxVx,$$

says the temperature *Tx* of a gas sample *x* is *identical to* some combination of *x*'s pressure *Px* and its volume *Vx*: that would be nonsense. All the gas law says is that *Tx*'s *value* is proportional to a function (the product) of the *values* of *Px* and *Vx*. Similarly, all a deterministic kinetic theory says is that *Tx*'s value is proportional to the value of a function (the mean) of the kinetic energies of *x*'s molecules. That theory may be false – it is – but not because it identifies two demonstrably distinct properties: it doesn't.

That is why our beliefs and desires differ from our brain states: they too are properties of different things. The former are properties of people (Strawson 1959, ch. 3.5), and the latter of our brains or, rather, of the varying congeries of brain cells on which, as we have seen, each of our varying beliefs and desires depends causally at any one time. And that is all Armstrong's theory of the mind need, and should, assert: not that a given belief or desire *is* a brain state, but that whether we have it or not is a two-valued function of the properties and relations of our brain cells, just as whether water boils is a two-valued function of its temperature and pressure. So understood, Armstrong's theory, like the kinetic theory of gases, may be false, but not because it asserts demonstrably false identities between properties of different things.

7 Beliefs and desires

Yet if beliefs and desires, if not identical to brain states, are still functions of them, the question asked in §6 re-arises: if brain states can have all the 'behavioural effects, perceptual causes and mutual interactions' of beliefs and desires, why postulate the latter at all?

The answer is that we need these mental states to provide truthmakers for truths about what we believe and desire. We need them even if Armstrong (e.g. 2003, 13) is wrong to hold that all truths need truthmakers, as I argue he is in

Mellor 2009, §8, where I endorse the 'moderate' truthmaker theory, of Heil 2000 and others, that

> only some truths, the primary ones, have truthmakers, while other truths and falsehoods are derivable from the primary truths by means of truth conditional semantics (Forrest & Khlentzos 2000, 3).

For even this moderate theory will still require contingent truths about what we believe and desire to have truthmakers, since they are not entailed by primary truths. That's because, with a few debatable Cartesian exceptions (like 'I exist'), for no contingent P, or fallible human x, is 'x believes P' or 'x desires P' a complete truth function of 'P'. My pub's being open does not entail that I believe it is, or believe it isn't, or that I want it to be open, or want it not to be; nor do any of those beliefs and desires follow from my pub's being shut.

More to the present point, no truths about what we believe or desire at any one time are entailed by truths about the brain states of which, at that time, they happen, contingently, to be functions. That is not just because those functions are contingent on biochemical laws. Suppose believing my pub's open makes me want to go there, even if in the end I stay at home. That causal link itself depends on my other beliefs and desires: for example, on my wanting a drink and believing my pub has the drink I want. So the belief that my pub is open is only what in §5 I called an INUS condition of the desire it causes, just as it is of the actions it causes. But then the brain state of which my belief is a function can also only be an INUS condition of the state of which the desire it causes is a function: the causal link between those brain states will be contingent on many other such states. And similarly for every other causal link between the brain states of which our constantly varying beliefs and desires are functions. But then which of these states is, at any time t, the one a given belief or desire is a function of will depend on which other brain states it is causally linked to at t.

This is why 'I believe my pub is open' and 'I want to go there' cannot be made true by my brain states: for no suitably causally connected values of B are they entailed by 'I am in brain state B'; and similarly for all other truths about what we believe and desire. Those truths can only be made true by our having those very beliefs and desires. And why should that not be what makes these truths about our beliefs and desires true, when every true 'x has mass Mx' is made true by x's having mass Mx? That after all we saw in §§3–4 to be more than a trivial application of §1's

(EP) 'P' is true iff P

since it embodies the striking and contingent fact that a single property of x, its mass Mx, gives x all the logically independent dispositions, to make different applied forces cause it to accelerate differently, which are entailed by the law that

$$Mx = Fx/Ax.$$

It is an equally striking and contingent fact about us, that a single property – a belief or desire – gives us so many logically independent dispositions, by making other beliefs and desires cause us to to act in so many different ways. That is what makes our beliefs and desires, and not any of our brain states, truthmakers for the truths, about what we believe and desire, which explain our behaviour. And while this is not a conclusion David Armstrong would endorse, it is still one I could not have reached without his work. For that, as for much else, I remain deeply indebted to him.

Bibliography

Armstrong, D. M. 1993. *A Materialist Theory of the Mind*, revised edn. London: Routledge.
Armstrong, D. M. 2003, "Truthmakers for Modal Truths". In Lillehammer, H. and Rodriguez Pereyra, G. (eds.), *Real Metaphysics: Essays in Honour of D. H. Mellor*. New York: Routledge, 12–24.
Armstrong, D. M. et al. 1996. *Dispositions: A Debate*, ed. T. Crane. New York: Routledge.
Ayer, A. J. 1946. *Language, Truth and Logic*, revised edn. Oxford: Oxford University Press.
Braddon-Mitchell, D., and Jackson, F. 2007. *Philosophy of Mind and Cognition: An Introduction*, 2nd edn. Oxford: Blackwell.
Carnap, R. 1936–7. "Testability and Meaning". In Feigl, H. and Brodbeck, M. (eds.), *Readings in the Philosophy of Science*. New York: Appleton-Century-Crofts (1953), 47–92.
Dyke, H. 2008. *Metaphysics and the Representational Fallacy*. London: Routledge.
Einstein, A. 1923. "The Foundation of the General Theory of Relativity", trans. W. Perrett and G. B. Jeffery. In Einstein. A. et al. *The Principle of Relativity*. London: Methuen, 109–64.
Foot, P. 1978. *Virtues and Vices*. Berkeley: University of California Press.
Forrest, P. & Khlentzos, D. (2000). "Introduction: Truth Maker and its Variants". *Logique et Analyse*, 43, 3–15.
Heil, J. (2000). "Truth Making and Entailment". *Logique et Analyse*, 43, 231–42.
Jackson, F. 1977. "Statements about Universals". In Mellor, D. H. and Oliver, A. (eds.), 1997, *Properties*. Oxford: Oxford University Press, 89–92.
Kripke, S. A. 1971. "Identity and Necessity". In Munitz, M. K. (ed.), *Identity and Individuation*. New York: New York University Press, 135–64.
Mackie, J. L. 1965. "Causes and Conditions". In Sosa, E. and Tooley, M. (eds.), *Causation*. Oxford: Oxford University Press (1993), 33–55.
Mackie, J. L. 1977. *Ethics: Inventing Right and Wrong*. Harmondsworth: Penguin.
Maxwell, J. C. 1873. *A Treatise on Electricity and Magnetism*. Oxford: Clarendon Press.

Mellor, D. H. 1974. "In Defence of Dispositions". In *Matters of Metaphysics*. Cambridge: Cambridge University Press (1991), 104–22.
Mellor, D. H. 2000. "The Semantics and Ontology of Dispositions". In *Mind, Meaning, and Reality*. Oxford: Oxford University Press (2012), 78–95.
Mellor, D. H. 2009. "Truthmakers for What?". In *Mind, Meaning, and Reality*. Oxford: Oxford University Press (2012), 96–112.
Newton, I. 1713. *Mathematical Principles of Natural Philosophy*, trans. A. Motte and F. Cajori. Berkeley: University of California Press (1962).
Quine, W. V. O. 1948. "On What There Is". In Id., 1953. *From a Logical Point of View*. Cambridge, Mass.: Harvard University Press, 1–19.
Ramsey, F. P. 1925. "Epilogue". In *Philosophical Papers*. Cambridge: Cambridge University Press (1990), 245–50.
Ryle, G. 1949. *The Concept of Mind*. London: Hutchinson.
Shoemaker, S. 1980. "Causality and Properties". In Mellor, D. H. and Oliver, A. (eds.), 1997, *Properties*. Oxford: Oxford University Press, 228–54.
Strawson, P. F. 1959. *Individuals: an Essay in Descriptive Metaphysics*. London: Methuen.
Tarski, A. 1944. "The Semantic Conception of Truth". In Feigl, H. and Sellars, W. (eds.), 1949. *Readings in Philosophical Analysis*. New York: Appleton-Century-Crofts, 52–84.

Francesco Orilia
Armstrong's Supervenience and Ontological Dependence

1 Introduction

In Armstrong's mature ontology, the one in which states of affairs and truthmaking gain center stage, there is a distinctive and pervasive appeal to a certain notion of supervenience. It is so defined: A *supervenes* on B iff (i) it is possible that B (the subvenient) exists and (ii) it is impossible that B exists and A (the supervenient) does not (Armstrong 1989 (CTP, hereafter), ch. 8; Armstrong 1997 (WSA, hereafter), 11).[1] The second clause can equivalently be put as follows: necessarily, if B exists, then A exists, i.e., the existence of B entails the existence of A, which suggests that supervenience is, in Armstrongian terminology, a *necessitation* relation: given the supervenience of A on B, we may say that A entails or necessitates B (WSA, 12, 92).

This way of speaking quite often invites a strict interpretation of these locutions, according to which there is Armstrong's world-view no ontological commitment to the existence of the supervenient. For example, he claims that "[t]he doctrine of the ontological free lunch rids us of superfluous entities" (WSA, 13). And thus it is not surprising to find philosophers who have interpreted thus strictly his "doctrine of the ontological free lunch," or *FL*, as we may call it in brief. However, typically, these philosophers have also pointed out that *FL* (so interpreted) leads to incoherence. For example, Oliver (1996, n. 30, p. 31) says: "Since supervenient entities exist and are not identical to the entities upon which they supervene, they must be an ontological addition." In a similar vein, Lowe (2011) remarks: "I wish to voice some concern about his notion of the 'ontological free lunch'. A free lunch is a *lunch*, and so something rather than *nothing*. But 'no addition of being' sounds very much like nothing to me." (For analogous perplexities, see, e.g., Melia 2005, 74 ff., and MacBride 2014, §2.1.)

---

[1] Although I have used, like Armstrong, the "s" of the third person singular in the verbs used for this definition, the variables "A" and "B" should be taken to range on both single entities and on pluralities, for Armstrong often applies supervenience to pluralities. For instance, as we shall see, he tells us that an internal relation supervenes on its relata or that an aggregate supervenes on its parts. Various notions of supervenience are often appealed to in much contemporary analytic philosophy, especially in philosophy of mind (McLaughlin and Bennett 2014), but they typically apply to properties, rather than to entities in general.

Francesco Orilia: University of macerata, email: francesco.orilia@unimc.it

These qualms originate from a less than charitable reading of what Armstrong has in mind, for after all he repeatedly makes it clear that he does not disavow an ontological commitment to the supervenient. For example, he tells us: "the second-class properties are not properties additional to the first-class properties. But it is to be emphasized that this does not make the second-class properties unreal. They are real and cannot be talked away" (WSA, 45). Analogously, he warns us that taking an internal relation to be supervenient "is *not* to say that ... it does not exist" (Armstrong 2004, 9).

Perhaps we should then look for an alternative, less strict, interpretation of FL. An option that may come to mind, explored by Keinänen (2008), is this *no distinctness proposal*: FL tells us that the supervenient is not distinct from the subvenient, where the distinctness in question is not necessarily numerical distinctness, but some other notion of "distinctness" deployed by Armstrong. This road is however highly problematic, as we shall see in §3, after the brief sketch of Armstrong's ontology of §2. Another line has been hinted by Correia (2005, 146) and more recently by Schaffer (2009, § 1.2) and Calemi (2013, 120 ff.): what Armstrong really means (or should mean) with FL is that the supervenient is ontologically dependent for its existence on more fundamental entities, namely its subvenient base; in other words, the supervenient exists *by virtue of*, or *because of*, the existence of the subvenient (For brevity's sake, I shall typically use "depends," "dependent" or "dependence" to express ontological dependence.) To be sure, this *dependence proposal* is also problematic, because in several cases in which Armstrong claims that some entities supervene on others, it is quite hard to admit that the supervenient depends for its existence on the subvenient. For example, Armstrong claims that the members of a set (or class) supervene on the set (and vice versa), whereas all supporters of ontological dependence agree that a set does not depend on its members, but it is rather the other way around (Correia 2005, 54, 138; Fine 1994, 4–5; Lowe 2010, §4, Schaffer 2009, 364: n. 22). Indeed, this example is typically offered by such supporters in order to provide an intuitive grasp of what dependence amounts to. Moreover, there are also cases in which we may want to claim an ontological dependence and yet there is no supervenience. Despite these difficulties, however, the dependence proposal may well shed new light into the ontological picture presented by Armstrong and I shall then explore it from §4 onward. This exploration will lead to a couple of suggestions regarding a problem with which Armstrong has struggled over the years, namely the nature of the link between universals and particulars in a state of affairs; and with that this paper will come to an end.

2 The basics of Armstrong's ontology

Following Mumford (2007, 183), we can distinguish in Armstrong's mature views, centered on states of affairs, an "official Armstrong philosophy," presented in detail in works such as CTP and WSA, and an alternative picture presented in *Truth and Truthmakers* (2004; T&T, hereafter). According to Munford (2007, 185), the crucial difference has to do with an "all-pervading presence of contingency" in the former, and a "major new kind of necessity in the world" in the latter. At bottom, this difference has to do with the wildly diverging ways in which states of affairs are understood in these two phases. In this work, I shall concentrate on the official Armstrong philosophy and in particular on its arguably most detailed and complete expression, namely the one found in WSA (although, when appropriate, there will be occasional references to other works of his, especially CTP).

Armstrong's ontology contemplates entities of three mutually irreducible basic categories, particulars, universals, and states of affairs (simply *states*, for brevity's sake). They all exist contingently, for they might have failed to exist. Universals can be either properties or relations and are all *in rebus*, i.e. they are instantiated (at some time or another) and cannot exist without being instantiated. When a property-universal F is instantiated (or exemplified) by a particular a, then there also exists another contingent entity, the state of affairs Fa. Similarly, if a relation R is instantiated by particulars in an appropriate number (and in a given order), say, a and b, then, contingently, there also exists the relational state of affairs Rab. These states are contingent not only because their constituents, the universal and the particular(s), might have failed to exist, but also because they might have failed to be so combined. For example, F might have been instantiated by b, rather than by a: Fb; and G by a rather than by b: Ga. For this reason, the world is a "world of states of affairs" and states should be counted as basic components of the world, additional to particulars and universals.

Armstrong also acknowledges in his ontological inventory properties and relations that are not universals, and thus are not among the fundamental entities (they will be reviewed in §6). All properties and relations, whether universals or not, are, at least in typical cases, multiply instantiated, they are "one over many." For example, let us suppose, the property H is instantiated by two different particulars, a and b. But universals have two further distinguishing features: they account for the resemblances and differences among particulars that we find in nature (WSA, 25) and contribute to the causal structure of the world, by bestowing on particulars powers to act and/or be acted upon (WSA, § 3.82).

3 The no distinctness proposal

Could not FL simply mean that the supervenient is numerically identical to the subvenient? This is most unlikely, for Armstrong (WSA, 12) explicitly tells us that "[s]ymmetrical supervenience yields identity." This surely suggests that, at most, the numerical identity of the supervenient and the subvenient can be assumed only if the supervenience in question is symmetric. But, as we shall see in detail, Armstrong admits many cases of asymmetric supervenience. Moreover, even when Armstrong claims that there is a symmetric supervenience and thus an identity between the subvenient and the supervenient, it is difficult to understand him as telling us that the identity in question is numerical. The paradigmatic example of symmetric supervenience offered by Armstrong has to do with aggregates: "the mereological whole supervenes upon its parts, but equally the parts supervene upon the whole" (WSA, 12). But then, if the parts and the whole are mutually supervenient, there would be *many* items, the parts, numerically identical to a *single* item, the whole, which seems absurd (Sider 2005, 15).

We have thus ruled out that there is no numerical distinctness between the supervenient and the subvenient. Armstrong however acknowledges other kinds of "distinctness" and lack thereof, and thus perhaps FL should be interpreted in terms of them. Let us see.

As we can gather from the previous section, it is crucial for Armstrong to focus on entities, x and y, such that the one can exist without the other, i.e., such that (i) it is possible that x exists and y does not, and (ii) it is also possible that y exists and x does not. When this is the case, x and y are, in Armstrongian terminology (which I shall follow consistently), *Hume distinct*, or, equivalently, *independent* of each other.[2] In sum, the Hume distinctness of two entities is their possible lack of co-existence, which does not rule out that they happen to co-exist (CTP, 22). Most characteristically, according to Armstrong, the states made up of universals and basic particulars, discussed in the previous section, are all Hume distinct; this is, says Armstrong (WSA, 1), "the Tractarian thesis of Independence." Could

---

[2] I prefer "Hume distinct" to "independent," so as to avoid a confusion with ontological independence. This use of the term "Hume distinct" is proposed by Armstrong in CTP, where he occasionally also resorts to "Hume independent" (CTP, ix, 41, 63). WSA does not contain "Hume" used as qualifier in this fashion, but Hume is cited therein while talking about "distinct existences" (WSA, 18). The use of "independent" as terminological variant of "distinct" occurs both in CTP and in WSA, though typically in talking about states of affairs. Armstrong occasionally says "wholly distinct" when he speaks, or should speak, of two independent entities (see, e.g., WSA p. 139, p. 265; CTP, 104), but as we shall see in a moment, he typically reserves another usage for this expression, and I shall stick consistently to this other usage.

FL simply men that the supervenient is not Hume distinct from the subvenient? Well, this lack of Hume distinctness is so trivially entailed by the definition of supervenience that all the emphasis that Armstrong places on FL makes no sense.

Armstrong also plays quite a bit with a notion of whole distinctness, and thus often tells us that certain entities are "wholly distinct." Following his typical use of this expression, let us say that x and y are *wholly distinct* iff they have no parts in common (CTP, 116; WSA, 18), where "parts" is used in a broad sense so as to speak both of parts of mereological wholes, sums or aggregates (Armstrong uses these expressions indifferently), and of constituents of complex entities such as states and conjunctive properties. Thus, for example, the leg of a table is not wholly distinct from the table, and two houses with a common wall are not wholly distinct. Similarly, the state Fa is not wholly distinct from the state Fb, and the conjunctive property F&G is not wholly distinct from the conjunctive property G&H. It should be clear that, if there is not whole distinctness, there cannot be Hume distinctness either. When there is lack of whole distinctness, Armstrong often deploys the term "partial identity."[3]

Could he not perhaps mean that the supervenient is not wholly distinct from the subvenient? This is how Keinänen (2008) interprets FL. This interpretation, however, is not very charitable. Keinänen all too easily shows that in almost every case in which Armstrong claims a supervenience, he is not justified in claiming that FL applies, simply because there is lack of partial identity between the supervenient and the subvenient. For example, an internal relation R that supervenes on x and y has no part in common with x and y.[4]

---

[3] Thus, for example, the leg of the table is partially identical to the table and the two houses with a wall in common are partially identical (WSA, 18); and such are the properties F&G and G&H (WSA, 31, 51) and the states of affairs Fa and Fb, or Ga and Fa (WSA, 140). I shall adhere consistently to this terminology, but it should be noted that Armstrong is not similarly consistent. For example, in WSA, 265, he uses "partial identity" for lack of Hume distinctness. Armstrong suggests (WSA, 14) that partial identity is a kind of loose identity. In addition to partial identity, he considers these two further examples of loose identity (WSA, 15): (i) in ordinary language we may say loosely say that two individuals share the same property, when in fact they instantiate two similar universals that strictly speaking are different; (ii) we may say that there is the same person at two different times, when in fact there are different temporal parts of a temporally extended object. These cases of loose identity play no role in Armstrong's account of supervenience and thus I shall neglect them.

[4] Hence, Keinänen sees FL as hardly defensible (unless it is seen as merely applying to mereological wholes).

4 Supervenience and ontological dependence

Let us then turn to the dependence proposal, the idea that FL is (or should be) a way of saying that the supervenient is ontologically dependent on the subvenient. This option is certainly intriguing and worth exploring, for it seems in line with Armstrong's insistence on (contingent) particulars, universals and states of affairs as somehow fundamental: everything else seems to depend on this basic level. But before delving into this, let us briefly characterize ontological dependence and fix some terminology (I shall rely on Schaffer 2009, §3.1).

When y is (*ontologically*) *dependent* (for its existence) on x, we can also say, with alternative terminology, that x *grounds* y (Schaffer's preferred way of speaking), that x is *prior* to y, that y is *posterior* to x (in all cases, the modifiers "ontologically" or "metaphysically" are usually inserted, but I allow for skipping them, for brevity's sake). Opinions diverge as to whether we should take ontological dependence as analyzable in more basic terms (Lowe 2010) or rather as a primitive notion that can be grasped by appropriate paradigmatic examples (Schaffer 2009). I tend to side with the latter option, but we can leave this open for present purposes. In any case, there seem to be logical constraints on the dependence relation. In the first place, something x can depend on more than one entity; in which case we can say that any of these entities is a partial ground of x. Moreover, it is typically assumed to be irreflexive, asymmetric[5] and transitive[6]. As we shall see, it is also useful to speak of the *generic* dependence that an entity x may have on *kinds* of entity. The idea here is that x is ontologically dependent on there being entities of a given kind K, although it does not matter to x which particular entities of kind K exist.

A thorny issue is whether ontological dependence must always be well-founded or not. If the former (Cameron 2008, Lowe 2010, Schaffer 2009), all chains of dependence must ultimately land on a fundamental entity (or plurality of entities), such that nothing further grounds it (this is the dominant view); if the latter (Correia 2005, 63; Orilia 2007, Gaskin 2008), there may be chains of dependence that go on indefinitely, without ever reaching a fundamental entity. There will be reasons to go back to the issue of well-foundedness or lack thereof.

In assuming that there is such a thing as ontological dependence, we come to view the ontologist as in charge of providing not only an inventory of entities and categories of entities, but also a hierarchical picture of reality, within which

---

5 But see Correia, 2005, 81.
6 When x depends on y, and there is no z in between such that x depends on z and z on y, we can say that x *directly* depends on y.

certain entities are more fundamental than others, with the former grounding the latter, a picture involving some explanation of why there is such a structure and how the less fundamental arises from what is more fundamental (Schaffer 2009). When Armstrong puts forward his own basic level of particulars, universals and states of affairs in the real world, he seems committed precisely to provide one such hierarchical picture of reality. In particular, it seems most appropriate to view Armstrong's "possible worlds" as ontologically dependent on the particulars and universals available in the real world. For such possible worlds are, as I understand Armstrong, sets (or conjunctions) of all the "state-of-affairs-like" combinations of the basic particulars and universals available in the real world (CTP, WSA, §19.25). Thus, for example, if there are in the real world the states of affairs Fa and Gb, but not the state of affairs Fb, there is nevertheless a state-of-affairs-like combination of F and b, which we may represent as "C(F, b)." Such combinations are, we may say, *merely possible* state of affairs, although Armstrong calls them simply *possibilities* (WSA, 160) (and does not represent them as I have suggested here). Armstrong would probably not be happy with this attribution of merely possible states of affairs to him, but they are, I think, implicit in his combinatorial theory of possibility.

Yet, the dependence interpretation of FL is also tortuous. First of all, the supervenience of A on B is certainly not a sufficient condition for the ontological dependence of A on B; one may think it is a necessary condition for this dependence, but even that can be questioned. It can thus at best be seen as a (fallible) indication of dependence; I shall dwell on such issues in the next two sections. Moreover, there are cases in which Armstrong claims a supervenience, but an ontological dependence does not seem plausible; I shall review these adverse cases in § 7.

5 Supervenience as neither sufficient nor necessary for ontological dependence

That the supervenience of A on B is not sufficient for the dependence of A on B is usually illustrated by recourse to the standard conception of sets. By its lights, once you are given, e.g., the objects x and y, there is by necessity the set {x, y}, with x and y as its sole members. But of course, if there is the set {x, y}, its members x and y cannot fail to exist. In sum, {x, y} supervenes on x and y and vice versa. Yet, intuitively, we want to say that {x, y} exists, because x and y exist, and not vice versa. Armstrong provides us with a further example of supervenience without dependence, when he tells us that the instantiation by a given object, a, of two different universals, P and Q, grants the existence of the conjunctive universal P&Q (also instantiated by a) (CTP, 113; WSA, 32).[7] In other words, P&Q supervenes on Pa and Qa. Yet, it seems wrong to say that P&Q depends for its existence on Pa and Qa, for P&Q could have existed even without Pa and Qa. For example, given the states Pb and Qb, we would also have P&Q. This suggests that conjunctive universals are generically dependent on the co-instantiation (by some object or other) of their conjuncts.

Alongside with sets, internal relations are often presented as paradigmatic examples of dependent entities. They may suggest that supervenience is at least a necessary condition for dependence. Consider for instance an internal relation such as being longer and suppose that there are two states of affairs such as x's being two meters long and y's being 2.1 meters long. Given the existence of these monadic states, there is by necessity the relational state consisting of x's being longer than y. Hence, x's being longer than y supervenes on the two monadic states. On the other hand, these two monadic states do not supervene on the relational state. For x's being longer than y might have existed, without the existence of the two monadic states. For example, it would have existed, if x would have been 2.3 meters long and y 2.2 meters long. This prevents us from saying that our two original monadic states exist by virtue of the existence of the relational state. Thus, one may think, the supervenience of A on B is necessary for the dependence of A on B; without supervenience, one might think, there cannot be dependence either.

Nevertheless, there may be reasons to think that some entity may be dependent on another, without supervening on it. Let us turn back to Armstrong's basic level. Even though Armstrong is not willing to admit it, there seems to be a hierarchy there, for particulars and universals appear to be more fundamental than states of affairs. After all, a state of affairs Fa seems to arise from its constituents F and a, which, in turn, by being its constituents, appear to be prior to it. This suggests that Fa depends for its existence on F and a (and not vice versa). Of course, according to Armstrong, they must be involved in some state of affairs or another, say in Fb and Ga, respectively, which suggests that there is a generic dependence of universals and particulars on states of affairs. Yet, it remains true that Fa does not supervene on F and a, for F and a could exist without Fa. Hence, it appears that the supervenience of A on B cannot in general be considered a necessary condition for the dependence of A on B, despite what the above example regarding

---

[7] On the other hand, even if the universals P and Q exist, there is no conjunctive universal P&Q, if P and Q are instantiated only by different objects (WSA, 31). For the reasons adduced by Armstrong in favor of conjunctive universals, see WSA, 32 and CTA, 113.

internal relations suggests. On the other hand, this case confirms that the supervenience of A on B is not a sufficient condition for the dependence of A on B. For of course F and a supervene on Fa: since they are its constituents, it is not possible that Fa exists, while F and a do not (we are assuming of course that Fa is possible).

For another example of dependence without supervenience, consider higher-order states of affairs. They arise from the instantiation of properties and relations by states of affairs or other properties and relations (WSA, 139, 196). Laws of nature are a case in point, as they involve relations between two universals (WSA, 197). Another example is provided by singular causation, understood as a relation that can link two states of affairs, say Fa and Gb, so that there is the state of affairs of Fa's causing Gb (WSA, 196). It seems appropriate to say that higher-order states of affairs depend for their existence on the instantiating universals or states of affairs, for the latter are constituents of the former. Thus, for example, the state of affairs of Fa's causing Gb depends on its constituents Fa and Gb. But the instantiating items might have existed without instantiating the property or relation that they happen to instantiate: Fa and Gb might have failed to be in a causation relation. Similarly, two universals connected in a law of nature might have failed to be so connected (the laws of nature could have been different). Of course, there may still be a generic dependence of state of affairs on higher-order states. Perhaps, one could argue, states of affairs cannot exist without being related to other states of affairs by causal connections or by connections of other sorts.

6 Cases of asymmetric supervenience

Even though supervenience is neither necessary nor sufficient for dependence, perhaps we can at least take the supervenience of A on B as an indication of the dependence of A on B, as internal relations suggest. Since dependence is asymmetric, this is easier when the supervenience is also asymmetric, for in that circumstance the direction of the dependence is clearly indicated. We shall now review many cases in which Armstrong claims an asymmetric supervenience that can be plausibly taken as indicating an ontological dependence.

Grounded Internal relations

We have already discussed relational states of affairs asymmetrically supervenient on monadic states of affairs. Let us simply record here that Armstrong is committed to such relational states, to the internal relations involved in them, and to their asymmetric supervenience on the relevant monadic states (CTP, Ch. 8, §II; 105, WSA, Ch. 6). Following Keinänen (2008, 55), we may call such internal relations *grounded*, so as to distinguish them from internal relations of another sort, to be considered in §7.

Second-class properties

Armstrong (1997, 44) distinguishes between properties that are universals and other properties that are not. The former are called *first-class properties*. The latter, by contrast, are called *second-class properties*.[8] Armstrong tells us that they are "perceptual properties, the properties of the 'manifest image'" and uses color-qualities by way of illustration (WSA, 45). As regards them, we are told that second-class properties are contingently instantiated by particulars and accordingly give rise to *second-class* states of affairs. The second-class states of affairs are seen by Armstrong as supervenient on first-class states of affairs, states of affairs involving only first-class properties, and, by the same token, the second-class properties are seen as supervenient upon the first-class properties (1997, 45–46).

Kinds

Armstrong (WSA, § 4.3, 65) considers that natural kinds are repeatables and thus have a *prima facie* right to be considered properties of things. On the other hand, they can be mutually exclusive and, traditionally, are linked to essentialism: if a thing is of a certain kind, it cannot exist, without being of that kind. Hence, as Munford (2007, 103) puts it, "they would impinge on combinatorial freedom." Kinds thus pose a dilemma to Armstrong. His way out is to view them as properties that supervene on the basic states of affairs.

---

[8] Despite a renewed emphasis on FL in this context (WSA, 45), Armstrong tells us that, since second-class properties are contingently instantiated by particulars, they give rise to an "increase of being over the particular that the property attaches to." He contrasts them with *third-class* properties, which instead are no increase of being. They are not, because, if true of an item, a, they are necessarily true of a. This is so, without there being a state of affairs that consists of the instantiation of a property by a, given that all states of affairs are taken to be contingent. The example provided by Armstrong is *identical to a*, which is necessarily true of a (WSA, 44). This distinction between a free lunch with an increase of being and one without is perplexing. But perhaps, once we interpret FL in terms of ontological dependence, it is not that important.

Relational properties

Armstrong (WSA, 92) admits relational properties of the sort *having R to a G*, which supervene on states of affairs such as Gy and Rxy, and of the sort *having R to x*, that supervene on states of affairs such Ryx.

Dispositions

Armstrong (WSA, 81–82) accepts dispositional properties as supervenient on laws of nature involving categorical properties; for example, a dispositional property D is asymmetrically supervenient on the law of nature (F&G)→H, where F and G are categorical properties and the arrow represents a relation of nomic necessitation.

The totality fact and negative facts

There is, according to Armstrong, a so-called totality state of affairs Tw, consisting of an 'allness' or 'being total' property, T, instantiated by the aggregate of all states of affairs, w (CTP, 94, WSA, 198). The existence of this state of affairs tells us that the states in w are all the states that there are. There might have been further states, however. For example, a universal F, not instantiated by the particular a, could have been instantiated by it. In this case we would have a different aggregate of all states of affairs, w', and thus a different totality state of affairs, Tw'. By taking advantage of the property T and the totality state of affairs, Armstrong thinks he has all he needs to provide truthmakers for negative truths. Suppose that b does not instantiate F, so that Fb is not in w and it is true that b is not F. That b is not F, according to Armstrong, is made true by the totality fact Tw, since its existence rules out the existence of Fb. Negative states of affairs, and the negative properties involved in them, such as being non-F, are then introduced as supervenient on the totality state of affairs (WSA, 200).[9]

---

[9] That one really needs totality and negative state of affairs in order to find truthmakers for negative and general truths may perhaps be questioned (Mellor 2008), but I shall not pursue this issue here. One may also wonder why supervenient negative states of affairs should be acknowledged, once we put things in terms of truhmaking, as Armstrong does at this juncture. For, if the totality fact is all we need to make a negative truth true, what role is left for the negative state of affairs? The same perplexity could be raised for other cases of supervenience. For example, one may say, and Armstrong quite often speaks in this way, that a truth such as that a is longer than b is made true simply by the relevant monadic states of affairs, say that a is two meters long and that b is

7 Cases of symmetric supervenience

When supervenience is symmetric, it is harder to take it as an indication of dependence; since dependence is asymmetric, we need reasons that help us decide which direction of the supervenience should be privileged. I shall now review various cases in which Armstrong claims that there is a symmetric supervenience and clarify what the direction of the dependence is. An alternative course, to be considered in the next section is this: in at least some cases of an alleged symmetric supervenience, there is really no such thing, because the alleged entities that give rise to it can be disposed of after all.

Strict Internal relations

In addition to grounded internal relations, Armstrong also admits (in Keinänen's terminology) *strict* internal relations, which are characterized by the fact that relational states of affairs involving them supervene directly on their relata (for example, one may say, the mere existence of 1 and 2 necessitates that 1 < 2). In this case, however, the supervenience is clearly symmetric, for of course, if Rxy exists, x and y cannot fail to exist. The standard and plausible view is that it is the relation that exists by virtue of (the natures of the) relata. A case in point is given by the making true relation, which, according to Armstrong is internal, as discussed extensively in David 2005. When the truthbearer p and the entity s (typically a state of affairs) are such that s makes p true, it happens that there is the state s's making p true, involving the internal relation of making true as connecting s and p. This state of affairs supervenes on s and p, and, equally well, p and s supervene on the state of affairs. But clearly we want to say that there is such a state of affairs, because there are s and p, and not the other way around.

Mereological aggregates

Armstrong repeatedly tells us that a mereological whole supervenes upon its parts and by the same token the parts supervene on the whole (WSA, 12). If so, we are left with no indication as to the direction of the ontological dependence, if there

---

one meter long. If so, what role is left for a supervenient state of affairs such as a's being longer than b? But the fact is that Armstrong appears to be committed to these supervenient entities and once they are accepted, we get the hierarchical structure outlined in this paper.

is any. What to do? Following Lowe (2010, §1), it makes sense to distinguish between (i) an organic whole such as a living being, or perhaps even a bicycle, that can survive the substitution or loss of (some of) its parts, and therefore is not dependent on them (although it may be *generically* dependent on some parts or others) and (ii) a mere unconnected aggregate of items, or a given quantity of matter, for which there is a dependence of the whole on the parts (see also Correia 2005, 82). Since Armstrong accepts a principle of unrestricted composition (WSA, 120), which even allows for an aggregate F+a, made up of a particular a and a property F (not instantiated by a), it seems that he has in mind the latter sort of whole. If so, he should say that the whole depends on the parts; it makes sense to say that, if there is such a thing as F+a, it exists because there are F and a, and it is certainly not the other way around. The aggregate x+y is after all not just x and y, but x and y *taken together*, as is often said. It thus exists, one could say, by virtue of the existence of x and y, for without them, there would be nothing to take together. However, it does not make much sense to say that x+y exists by virtue of the existence of x and y. But should we really admit this unrestricted composition? We shall come back to this.[10]

Sets

As we saw above, it is typically assumed that sets supervene on their members and vice versa. Armstrong agrees with this, although he has his own idiosyncratic accounts of multi-membered sets as mereological aggregates of singletons, which in turn are viewed as somewhat special states of affairs (WSA, ch. 12).[11] We also saw that sets are paradigmatically taken to be dependent on their members. Given what we said above about aggregates, Armstrong could accept this unproblematically for multi-membered sets. As regards singletons, if they are states of affairs, their dependence on their solitary members appear to presuppose the idea that a state of affairs Fa depend (inter alia) on a for its existence. I suggested above that Armstrong should accept this. However, just as for unrestricted aggregates, I shall consider below if a commitment to sets (whether understood *à la* Armstrong or not) is appropriate.

---

[10] An interesting issue that I cannot tackle here is how Armstrong views organic wholes and their relation to the parts and matter that constitute them.
[11] There are various complications and difficulties in this approach (CTP, 191 ff.), which I cannot consider here.

Conjunctive states of affairs

Armstrong explicitly commits himself to conjunctive states of affairs (WSA, 35; see also CTP, 45). The conjunction at work in conjunctive states of affairs seems to be nothing more than mereological fusion (WSA, 35, 122). In other words, given any states of affairs $S_1, \ldots, S_n$, there is corresponding state $S_1 + \ldots + S_n$ (although it may be more intuitive to represent this as the state $S_1 \& \ldots \& S_n$). Conjunctive states of affairs are *molecular*, says Armstrong, and he considers whether there are other molecular states of affairs, e.g., negative and disjunctive ones, but he rules it out and proclaims (WSA, 35) that conjunctive states of affairs are the only molecular states of affairs.[12]

Since Armstrong treats conjunctive states of affairs as aggregates, he views them, like other aggregates, as symmetrically supervenient on their parts. Thus, for example, as he sees it, the state Fa & Gb is supervenient on Fa and Gb, and vice versa. Once more, then, supervenience does not give us an indication as to the direction of the dependence relation. However, in line with what we said above about aggregates in general, we should take conjunctive properties to be dependent on their conjuncts.

"Hyper-total" states of affairs

Let us go back to the totality state of affairs Tw. Clearly, it is a further state not comprised in w, which thus in a sense is not total after all. By adding it to w, we obtain a new aggregate w^1, which seemingly, should instantiate T, so that we get a further totality state of affairs Tw^1, of (higher) level 1 (we may take Tw to be of level 0 and represent it as Tw^0). And we can go on *ad infinitum* with totality states of ever increasing levels. According to Armstrong, however, this regress is harmless, because these further totality states are supervenient; once the first one is given, they are all necessitated (WSA, 198). It should be noted however that the supervenience is symmetric. For surely if Tw^1 exists, Tw must also exist. Similarly, for any higher level n, if Tw^n exists, so must Tw^{n-1}. Thus, to capture what Armstrong has in mind, it seems we must specify a direction for the dependence relation: for each level n, Tw^n exists by virtue of the preceding totality state, rather than *vice versa*.

---

12 Disjunctive and negative states of affairs are admitted, but not at the fundamental level.

8 Instantiation and Bradley's regress

Armstrong moves to the new ontological picture of T&T essentially because of his perplexities regarding the nature of instantiation, the tie that puts together universals and particulars so as to generate states of affairs. Out of dissatisfaction with previous attempts, in T&T Armstrong comes to view the instantiation of a universal by a particular as a partial identity of the particular with the universal. This is a difficult doctrine, which I have no room to discuss here. What I rather want to do is to examine two possible ways to come to terms with instantiation, which, if accepted, may defuse the temptation to embrace the complications of a theory of instantiation as partial identity and the shift of ontological paradigm that comes with it. After all, Armstrong himself was not content with it, since he does not seem to endorse it in the final survey of his views (2010).

Armstrong's problems with instantiation have to do with Bradley's regress. As a reaction to it, he ends up proposing in WSA that there is really no need to postulate a *fundamental* tie of instantiaton,[13] since it is the state of affairs itself that holds its constituents together in a non-mereological form of composition (WSA, 118).[14] Taken literally, this seems incoherent, for it looks as if the state of affairs should do something, putting its constituents together, before it can even come to exist as the complex entity that it is. But there may be a more charitable way of reading Armstrong. Perhaps, he is simply advancing the idea that the complexity of states of affairs is an ontologically brute fact, something that cannot be further explained by appealing to instantiation or anything else (Orilia 2007, 153): there are states of affairs, as complexes with their universals and particulars as constituents, and this is the end of the story.

There is an obstacle to this approach however, which has been pointed out by Vallicella (2002, 235–236). In addition to a state of affairs Fa, Armstrong also acknowledges another entity with the same constituents, namely the aggregate F+a. If they are different entities, then there must be something that differentiates them. But if they have the same constituents, what is it? If one could appeal to instantiation, one could answer that the difference is given by the fact that Fa exists by virtue of the instantiation of F by a. But we are assuming that we cannot take advantage of instantiation. Given this, we can perhaps follow another road.[15] Namely, we could reject unrestricted composition and deny that there are

---

[13] He then suggests that instantiation can be taken to be supervenient (WSA, 119).
[14] The composition is non-mereological, because states of affairs, contrary to mereological sums, can be Hume distinct, without being wholly distinct (WSA, 118–119, 140).
[15] I have explored this road in Orilia 2004, § 7.2 and 2014, § 3.2.

such aggregates as F+a. We saw however that Armstrong reconstructs sets in terms of aggregates and, one may say, we need set theory for mathematics and science (WSA, 187). If we get rid of aggregates, don't we then need sets as irreducible entities? But then again we get a set {F, a} that appears to have the same constituents as the state Fa. At this point, however, one can appeal to Russell's idea that all talk of sets can be reconstructed in terms of properties and relations, the so called "no-class theory of classes" (Landini 2011, 115–124). Problems are not over, however, because Armstrong admits relational states of affairs, Rxy and Ryx, with R non-symmetric, that are different despite having the same constituents (WSA, 140). If states of affairs are ontologically brute, it seems difficult to account for the difference between Rxy and Ryx, given that they have the same constituents. A way out that does not appeal to an instantiation tie is an account of relational states according to which there really are no states of affairs that differ while having the same constituents. If so, Rxy and Ryx differ because they have (despite appearances) different constituents. Let me briefly sketch a possible approach of this sort: the state Rxy really is a relation of two monadic states of affairs, say, Fx and Gy. In other words, it is R(Fx, Gy). Similarly, Ryx is R(Fy, Gx). If so, the two states have different constituents: R, Fx and Gy in one case, and R, Fy and Gx, in the other case. (This is in line with what is said in WSA, pp. 120–121, about conjunctive states of affairs). To illustrate, the state consisting of x's loving y, Lxy, would be analyzed as involving properties such as being an agent, A, and being a patient, P, instantiated by x and y, respectively: Ax, Py. So that Lyx would turn out to be: L(Ax, Py).[16]

The second way to come to terms with instantiation that I want to consider is what I have called *fact infinitism*. It is based on the idea that, rather than worrying about Bradley's regress, we take it on board as the key to explain how universals and particulars succeed in being tied together in states of affairs as entities that differ from mere aggregates. To illustrate, we say that there is the state of affairs Fa, something different from the mere aggregate F+a, because a instantiates F, so that there is a further state of affairs IFa, where I is instantiation. Of course, this requires an explanation of the existence of IFa and its diffference from I+F+a, which leads us to a higher-order instantiation I' and to a further state I'IFa; and so on, *ad infinitum*. This generates an endless chain of ontological dependence that never touches a final ground: Fa depends on Fa, which depends on IFa, which

---

[16] Agent and patient are two among many other *onto-thematic roles*, as I have called them. They can be appealed to in various ways in order to account for relatedness. Here I am outlining one possible way. See Orilia 2011 for a detailed discussion of other alternatives.

depends on I'IFa, etc. In other words, we have to give up well-foundedness (Orilia 2007; see Maurin 2015 for a critical discussion).

Could Armstrong ever accept this? Well, it seems that, once we project ontological dependence into his framework, he should not be committed to well-foundedness. So perhaps he could consider fact infinitism after all, even though he may judge it ontologically extravagant. To see why Armstrong should reject well-foundedness, we need turn to his suggestion that, for all we know, all universals may turn out to be, on empirical grounds, conjunctive (or, more generally, structural), all the way down *ad infinitum* (CTP, 113; WSA, 33, 123). Now, it certainly makes sense to say that, just as a set or sum is dependent on its members or parts, a conjunctive universal, P&Q, is dependent on its conjuncts. But then a universal that is conjunctive *ad infinitum* sets up an infinite chain of dependence with no ultimate ground: if P is conjunctive *ad infinitum*, it depends on a certain conjunct P_1, which in turn depends on another conjunct P_2, and so on; P is, we may say, *gunky*. In sum, if we are to understand FL in terms of ontological dependence, as we are trying to do, it seems we should enroll Armstrong in the party of those who do not take ontological dependence as necessarily well-founded. Schaffer 2010 argues from the empirical possibility of *gunks*, objects made up of smaller and smaller parts *ad infinitum*, to *priority monism*, the thesis that the cosmos is a whole on which everything else, *qua* part, is dependent. But this argument presupposes that the well-foundedness of dependence is taken for granted. However, if we rather take for granted the plausible idea that a complex such as P&Q is dependent on its parts and not vice versa (after all, according to Armstrong, as we have seen, P&Q might fail to exist even if P and Q exist), the possibility of *gunky* universals should rather lead us to question the well-foundedness of dependence. Similarly, well-foundedness should be questioned, given the possibility of gunks, or the possibility of appealing to fact infinitism to account for the relatedness of universals and particulars that brings about states of affairs.

Bibliography

Armstrong, D. M. *A Combinatorial Theory of Possibility*. Cambridge: Cambridge University Press.
Armstrong, D. M. 1997. *A World of States of Affairs*. Cambridge. Cambridge University Press.
Armstrong, D. M. 2004. *Truth and Truthmakers*. Cambridge: Cambridge University Press.
Armstrong, D. M. 2010. *Sketch for a Systematic Metaphysics*. Oxford: Oxford University Press.
Beebee, H. and Dodd, J. (eds.). *Truthmakers: The Contemporary Debate*. Oxford: University Press.
Calemi, F. F. 2013. *Le radici dell'essere*. Roma: Armando.

Cameron, R. 2008. "Turtles All The Way Down: Regress, Priority And Fundamentality". *Philosophical Quarterly*, 58, 1–14.

Correia, F. 2005. *Existential Dependence and Cognate Notions*. Munich: Philosophia Verlag.

David, M. 2005. "Armstrong on Truthmaking". In Beebee and Dodd J. 2005, 141–159.

Fine, K. 1994. "Essence and Modality". In Tomberlin, J. E. (ed.), *Philosophical Perspectives 8: Logic and Language*. Atascadero, CA: Ridgeview Publishing Company.

Gaskin, R. 2008. *The Unity of the Proposition*. Oxford: Oxford University Press.

Keinänen, M. 2008. "Armstrong's Conception of Supervenience". *Acta Philosophica Fennica*, 84, 51–61.

Landini, G. 2011. *Russell*, London: Routledge.

Lowe, E. J. 2010. "Ontological Dependence". In Zalta, E. N. (ed.), *The Stanford Encyclopedia of Philosophy* (Spring 2010 Edition), URL =<http://plato.stanford.edu/archives/spr2010/entries/dependence-ontological>.

Lowe, E. J. 2011. "D. M. Armstrong. Sketch for a Systematic Metaphysics". *Notre Dame Philosophical Review*, 2011.01.19, URL = <http://ndpr.nd.edu/news/24578-sketch-for-a-systematic-metaphysics>.

MacBride, F. 2014. "Truthmakers". In Zalta, E. N. (ed.), *The Stanford Encyclopedia of Philosophy* (Spring 2014 Edition), URL = <http://plato.stanford.edu/archives/spr2014/entries/truthmakers>.

Maurin, A.-S. 2015. "States of Affairs and the Relation Regress". In Galluzzo, G. and Loux, M. J. (eds.), *The Problems of Universals in Contemporary Philosophy*, Cambridge: Cambridge University Press, 195–214.

McLaughlin, B., and Bennett, K. 2014. "Supervenience". In Zalta, E. N. (ed.), *The Stanford Encyclopedia of Philosophy* (Spring 2014 Edition), URL = <http://plato.stanford.edu/archives/spr2014/entries/supervenience/>.

Melia, J. 2005. "Truth-making without Truth-makers". In Beebee and Dodd, 67–84.

Mellor, D. H., 2008. "Truthmakers for What". In Dyke, H. (ed.), *From Truth to Reality: New Essays in Logic and Metaphysics*, London: Francis & Taylor.

Mumford, S. 2007. *David Armstrong*, Stocksfield: Acumen.

Oliver, A. 1996. "The Metaphysics of Properties". *Mind*, 105, 1–80.

Orilia, F. 2004, "States of Affairs and Bradley's Regress: Armstrong Versus Fact Infinitism," unpublished ms. available at URL = <http://www.academia.edu/13245772/States_of_Affairs_and_Bradleys_Regress_Armstrong_Versus_Fact_Infinitism>.

Orilia, F. 2007. "Bradley's Regress: Meinong vs. Bergmann". In L. Addis, G. Jesson, and E. Tegtmeier, *Ontology and Analysis, Essays and Recollections about Gustav Bergmann*, Frankfurt: Ontos Verlag, 133–163.

Orilia, F. 2011. "Relational Order and Onto-thematic Roles". *Metaphysica*, 12, 1–18.

Orilia, F. 2014. "Particolari, proprietà, relazioni e stati di cose," in Fraisopi, F. (ed.), *Ontologie. Storie e prospettive della domanda sull'ente*, Mimesis, Milano-Udine, 2014, pp. 469–499 (apart from minor revisions, this coincides with a version originally published online (June 16, 2009) in URL = "http://www.giornaledifilosofia.net"www.giornaledifilosofia.net).

Schaffer, J. 2009. "On What Grounds What". In Manley, D., Chalmers, D. J., and Wasserman R. (eds.), *Metametaphysics: New Essays on the Foundations of Ontology*. Oxford: Oxford University Press, 347–383.

Schaffer, J. 2010. "Monism: The Priority of the Whole". *Philosophical Review*, 119, 31–76.

Sider, T. 2005. "Another Look at Armstrong's Combinatorialism". *Noûs*, 39, 680–696.

Vallicella, W. F., 2002. "Relations, Monism, and the Vindication of Bradley's Regress". *Dialectica*, 56, 3–35.

Paolo Valore
Naturalism as a Background Metaphysics

At present, Naturalism appears to be the dominant perspective in philosophy, with a growing trend toward anti-Naturalism.[1] Still, labeling a contemporary philosopher as "Naturalist" or "anti-Naturalist", without other clarifications, does not seem very informative.

Quine was a Naturalist inasmuch he dismissed any philosophical epistemology opposed to and distinct from the scientific inquiry. If your ideal model of science is physics, Naturalism entails Physicalism, but this is not necessarily so. This methodological Naturalism, that we could also call "Scientism", assumes no superior arbiter in knowledge related questions than science itself, but does not, as such, offer positive insights about the ultimate nature of reality or the *desiderata* of our epistemology: "The science game is not committed to the physical, whatever that means. [...] Even telepathy and clairvoyance are scientific options, however moribund" (Quine 1990, 20-21).

We may also claim Naturalism to be the project to eliminate any normative component not only in our account of knowledge, such as in Goldman 1986, but also in our account of action, e.g. in some projects of meta-ethics.[2] This Naturalism has its own agenda that is not, as such, prescribed by any particular science.

Lastly, we may consider Naturalism the reductionist project known as "eliminative materialism", according to which certain entities, such as mental states, are to be reduced to certain others: for instance, projects of naturalization of "mind" in terms of what is legitimately admitted by neurosciences. This Naturalism can be found in project of Naturalization such as that of the Churchlands (Churchland P.M. 1988; Churchland P.S. 1987). Naturalization is, basically, the project of bringing back to Nature what was supposed to possess a special essence. Here, Naturalism assumes that reality has a certain intrinsic nature and entails an ontology.

If there is a common, core assumption in different versions of Naturalism, this amounts to the denial of anything "supernatural", understanding "supernatural" as a method, a field or an entity that goes beyond Nature. Clearly, this ends up being circular, if we need to understand "Nature" as what a Naturalistic perspective acknowledges, and "Naturalism" as the denial of anything that is not Nature.

---

1 For a presentation of the debate, cf. Macarthur-De Caro 2004.
2 A classic example is Blackburn 1998, Chap. 3: "Naturalizing Norms".

Paolo Valore: University of Milan, email: paolo.valore@unimi.it

1 Understanding Armstrong's naturalistic position

According to Armstrong (1978a), "Naturalism" is "the doctrine that reality consists of nothing but a single all-embracing spatio-temporal system" (p. 149). Slightly rephrased, "Naturalism" is "the contention that the world, the totality of entities, is nothing more than the spacetime system" (Armstrong 1997, 7). The two definitions space about twenty years apart: given that they state essentially the same content, we may assume that "Naturalism" has a fixed meaning throughout Armstrong's philosophical itinerary.

The endorsement of Naturalism stands as a philosophical general background of Armstrong's research and it is rarely discussed or presented, at least explicitly. It comes together with two other general assumptions: Physicalism, which assumes that it is natural science, and physics in particular, "that gives us whatever detailed knowledge we have of the world" (Armstrong 1997, 5), and Factualism, which represents the core ontological position, according to which the simplest units of the universe are states of affairs or "Tractarian facts".

Clearly, Physicalism is linked, if not eventually identical, to Materialism: the world must be explained exclusively in terms in physics because "the only particulars that the spacetime system contains are physical entities governed by nothing more than the laws of physics" (Armstrong 1997, 6). At the end, it is because of what exists that we have to be physicalist; therefore, the epistemological obligation is connected to the ontological thesis. In turn, being tightly connected with the two other general assumptions, Naturalism might be seen as a way to paraphrase one or the other of the two. Armstrong himself seems in many ways to promote an overlapping. For instance, calling at times Naturalism "an ontological doctrine" (Armstrong 1997, 5), while at other times he is clear that "if we define ontology or 'first philosophy' as the most abstract or general theory of reality, then it seems that neither Materialism nor even Naturalism is an ontology" (Armstrong 1978a, 261) and that "one can be a factualist without being a Naturalist" (Armstrong 1997, 5). Another perplexing overlapping arises when Physicalism is told to be "a sub-species of Naturalism", while Naturalism and Physicalism are presented as two different doctrines (*ibid.*). Nonetheless, it seems to me that recalling the similarity of the Naturalist thesis with an ontology just serves the purpose to distinguish it from a mere epistemic conception, while distancing the Naturalist thesis from a proper ontology of state of affairs and putting it closer to Physicalism serves the purpose to distinguish it from an ontology as such.

We may say that Naturalism expresses the metaphysical framework, while factualism states the general ontology and physicalism the most general epistemological presupposition. What is shared is the unified perspective, according to which there is a univocal account of reality (metaphysics), of the ultimate constituents of being (ontology) and of the essential nature of what is real, to be known in terms of physics (epistemology).

2 The under-determination of the thesis

The Naturalistic conception just expressed is essentially under-determinated, since it allows for so many further questions that are to be left without any answer.

Since we *know* that the spacetime is a structured set of spacetime points, we may be tempted to assert *this* knowledge as part of our statement about the nature of reality. Naturalism would then import at least a minimalist ontology, for instance a minimalist account of the space-time system such that there must be as a minimum spacetime points as the most fundamental building blocks of the world. And yet, with the Naturalistic thesis, that is what I want to make clear here, we cannot even assert this knowledge: "Notice that the thesis of Naturalism, as it is understood here, is not committed to the view that space and time, or even spacetime points, are ontologically fundamental" (Armstrong 1997, 6).

Such a lack of a more grained definition of the space-time system is connected to the epistemological assumption of Armstrong's philosophy. Physicalism imposes that any further inquiry about the nature of the space-time system, i.e. of reality, should be dismissed as a non-philosophical question and acknowledged as a physical, *a posteriori* question. Naturalism is a philosophical matter; the nature of spacetime is not. "Science has provided us with a method of deciding disputed questions", Armstrong once said, referring to the nature of mind (Armstrong 1980). And the same could be claimed here: there is no room for philosophical creativity in assessing the nature and properties of space and time.

In telling us what the world is, Naturalism also sets up boundaries within which philosophy may raise questions. Nonetheless, this is not all that we, as philosophers, can say.

3 The negative content

If we want to escape a quasi-trivial position, we should avoid to express Naturalism as the position such that "reality consists of anything that/all that is in the spatio-temporal system". It is hard to imagine a philosophical doctrine, aside from a radical form of idealism or immaterialism, which denies that what is in the spatio-temporal system is real, whatever that means. That reality consists of anything that/all that is in the spatio-temporal system does not rule out that it might consist of something else too. Armstrong statement of Naturalism is, on the contrary, that "reality consists of *nothing but* a single all-embracing spatio-temporal system".[3] The world consist of the space-time system and *nothing else*.

Armstrong calls the existence of the space-time system the "positive component".[4] But because of the under-determination of the thesis and contrary to what might seems at a first sight, the statement that the world is such-and-such, and nothing more than such-and-such, informs us more about what the world is *not* than what the world is. It prevents us from acknowledging additional entities that show some other different nature, while leaving unresolved the question of the nature of what there is.

We know that Naturalism cannot offer an ontology (this will be delivered to Factualism) but it does give us a *negative ontology*: it tells us that are no Cartesian minds, private visual and tactual spaces, angelic beings and God (cf. Armstrong 1978b, 127), no numbers and universals, which exist and transcend spacetime (cf. Armstrong 1997, 5). Naturalism tells us nothing about what is to be included in the inventory of the world but tell us a lot about what does not exist.

4 The positive content

If there is a positive content in Naturalism as such, it doesn't seem to be the existence of the spacetime in itself, but rather the univocal account that implies a unified conception of space-time, such that there is just *a single* space-time system. The statement that reality is to be reduced to space and time would be compatible with the assumption of a plurality of space-time systems. This is not the position discussed here, for Armstrong explicitly rejects such possibility.

The metaphysical, a priori assumption that there is a single space-time system, cannot be discarded, not even on the basis of a physical result. Consider the case of the multi-worlds interpretation of quantum physics: such interpretation does not conflict with the unified account of space-time. On the contrary, this should be counted "as a theory of a single space-time, but one of an unorthodox shape" (Armstrong 1989, 17, note 1).

---

3 Armstrong 1978a, 149.
4 Armstrong 1997, 8.

When connected to the ontological factualist thesis, Naturalism tells us that there is single collection of states of affairs. In fact, if reality consists of nothing but a single all-embracing spatio-temporal system and if "the spacetime system is identical with a certain set or aggregate of states of affairs" (Armstrong 1999, 82), then reality consists of nothing but a single set or aggregate that is the totality of states of affairs. "The states of affairs, organized as they are organized, [...] constitute the whole of reality".

So, we may say that Naturalism is mainly a negative thesis with the positive assumption that there is a unified system of reality. The positive content is extremely tiny and yet, metaphysically, relevant. We could say that in *mundo non datur hiatus, non datur saltus, non datur casus, non datur fatum*.

5 Is the a priori back?

In dismissing other traditional, non-Naturalistic conceptions of reality, Armstrong does not spend much time considering such views, arguing that "the arguments used to establish them are all a priori. [...] as an Empiricist, I reject the whole conception of establishing such results by a priori argumentation" (1978a, 262).

If a priori argumentation is to be rejected, it does not seem clear how we could justify the assumptions imported by Naturalism with some a posteriori evidence. In case of particular, selected cases of supernatural entities it might be the case that "the fundamental objections to postulating these entities are scientific and even observational" (Armstrong 1978b, 127), i.e. completely a posteriori. But this cannot be true of other cases and, most importantly, cannot be the basis of a claim about what we have to consider Nature as a systematic whole.

It seems to me more convincing the acknowledgement that

> we have to accept [...] that straight refutation (or proof) of a view in philosophy is rarely possible. What has to be done is to build a case against, or to build a case for, a position. One does this, usually, by examining many different arguments and considerations against and for a position and comparing them with what can be said against and for alternative views. (Armstrong 1989, 18)

This cost-benefit analysis is directed by some meta-theoretical criteria, that guide our evaluation, but this second level criteria cannot be justified with the appeal to something else and, for sure, cannot be justified a posteriori. For instance, we may argue that, if the gain is great, it might be justified the subordination of the general ontology to the demands of the epistemological theory (cf. Armstrong 1973, 103).

But it is unclear which is the general criterion, considering that, other times, we might judge more justified the subordination of our epistemological theory to the demands of our general ontology.

Certainly, we may argue for the completely reasonable claim that we should have and we do have scientific evidence that what exist is such and such. A proper ontology ought to be justified a posteriori, by means of the scientific enterprise. Accordingly, any way to establish results by a priori argumentation should be rejected by any Empiricist. But when we come to the completely different thesis of the unity of the world or the system of Nature, or the thesis that such unity is systematic interconnection of physical laws, the choice of which higher level norms and standards are to be used to evaluate our thesis cannot be considered metaphysically innocent.

Drawing the boundaries of Nature, Naturalism is the premise and guideline that makes the refusal of any a priori and "supernatural" position even possible in philosophical terms. Yet, as a premise and a guideline, it is itself an a priori assumption, *as if* we could legitimately make one. The unified conception of Nature as systematically ordered by the laws of physics is, as such, outside and prior to the picture of the world that this conception permits to build.

I am *not* saying that there is a vicious circle nor am I looking condescendingly to such metaphysical assumption. I just think that we need to be aware that this situation is inevitable and we just need to face our believes for what they are.

We need to believe. We have to assume that, at least on the background and at least considering that our saying is an *as if* saying, a statement such as "the totality of entities is nothing more than the spacetime system" is both true and informative. With Kant, we could say that this belief is a necessary presupposition, without which our scientific picture of the world wouldn't be possible. Still, this presupposition is a belief and not proper knowledge of any kind and it is a priori and not found in any of our investigation of Nature. As Armstrong once said (1999, 82): "In the fields of philosophy and religion there is no knowledge. We can only know what our beliefs are".

Bibliography

Armstrong, D. M. 1973. *Belief, Truth, and Knowledge*. Cambridge: Cambridge University Press.
Armstrong, D. M. 1978a. "Naturalism, Materialism and First Philosophy". *Philosophia*, 8, 261–276.
Armstrong, D. M. 1978b. *Universals and Scientific Realism, Vol. 1, Nominalism and Realism*. Cambridge: Cambridge University Press.

Armstrong, D. M. 1980. *The Nature of Mind and Other Essays*. Brisbane: University of Queensland Press.

Armstrong, D. M. 1989. *A Combinatorial Theory of Possibility*. Cambridge: Cambridge University Press.

Armstrong, D. M. 1997. *A World of State of Affairs*. Cambridge: Cambridge University Press.

Armstrong, D. M. 1999. "A Naturalist Program: Epistemology and Ontology". *Proceedings and Addresses of the American Philosophical Association*, 73(2), 77–89.

Blackburn, S. 1998. *Ruling Passions: A Theory of Practical Reasoning*. Oxford: Claredon Press.

Churchland, P. M., 1988. *Matter and Consciousness*, revised edition. Cambridge (Mass.): MIT Press.

Churchland, P. S., 1987. "Epistemology in the Age of Neuroscience". *Journal of Philosophy*, 84, 544–553.

Goldman, A., 1986. *Epistemology and Cognition*. Cambridge (Mass.): Harvard University Press.

Quine, W. V. O. 1990. *Pursuit of Truth*. Cambridge (Mass.): Harvard University Press.

Macarthur, D., De Caro, M. 2004. *Naturalism in Question*. Cambridge (Mass.): Harvard University Press.

Index

Anjum, R. 163, 168, 171, 172
Aquinas, St. T. 168
Aristotle 62, 118, 127
Audi, P. 59
Ayer, A.J. 219

Balashov, Y. 139
Ballarin, R. 178
Baxter, D. 17, 22, 122
Bealer, G. 44
Beebee, H. 210, 211, 216
Bennett, K. 233
Benovsky, J. 145
Bergmann, G. 105–108, 129
Bhaskar, R. 162, 163
Bird, A. 27, 193, 198, 203
Blackburn, S. 253
Borghini, A. 8, 177, 187
Boyce, K. A. 55
Brännmark, J. 102
Braddon-Mitchell, D. 146, 157, 226
Bradley, F.H. 19
Butchvarov, P. 7, 105, 110–115, 129
Bynoe, W. 21

Calemi, F.F. 5, 42, 102, 234
Cameron R. 177
Cameron, R. 238
Campbell, K. 15, 48, 75, 80, 81
Carnap, R. 224, 225
Carroll, J. W. 98
Cartwright, N. 170, 193
Chisholm, R. 105, 150
Choi, S. 193
Churchland, P.M. 253
Churchland, P.S. 253
Cocchiarella, N. B. 44
Copeland, B.J. 178
Correia, F. 21, 26, 27, 234, 238, 245
Cumpa, J. 7, 108, 127, 128, 133, 135, 137

Daly, C. 212–214
Daniels, P. 141, 156

Darby, G. 177
David, M. 209, 244
De Caro, M. 253
Denkmayr, T. 74
Devitt, M. 13–17, 86, 214
Dodd, J. 18, 210, 211, 216
Dretske, F. 96
Dummett, M. 217
Dyke, H. 219, 220

Eagle, A. 152–154
Efird, D. 177
Einstein, A. 225
Ellis, B. 163, 193

Fara, M. 193
Feser, E. 163, 165, 166
Fine, K. 21, 26, 234
Foot, P. 219
Forrest, P. 229
Frege, G. 106, 113, 114

Garcia, R. K. 88
Gaskin, R. 238
Girle, R. 178
Goldman, A. 253
Groff, R. 163
Grossmann, R. 7, 105–108, 120, 121, 126–129

Hanfling, O. 162
Harré, R. 162, 163
Haslanger, S. 140, 145
Hawley, K. 139
Heil, J. 97, 229
Heller, M. 139
Hochberg, H. 92, 93, 95, 96
Horwich, P. 208
Hume, D. 161, 162, 169

Jackson, F. 220, 226

Künne, W. 42, 43
Keinänen, M. 234, 237, 242, 244

Khlentzos, D. 229
Kripke, S.A. 227, 228

Labossiere, M.C. 94
Landini, G. 248
Lando, G. 177, 187
Lewis, D.K. 17, 86, 98, 139, 162, 169, 177, 184
Liggins, D. 216
Loux, M.J. 43, 88
Lowe, E.J. 26, 140, 194, 202, 208, 212, 216, 233, 234, 238, 245

Macarthur, D. 253
MacBride, F. 48, 233
Mackie, J.L. 165, 219, 225
Madden, E. H. 162, 163
Marmodoro, A. 163, 168
Martin, C.B. 97, 161, 193, 198, 199
Maurin, A.-S. 6, 48, 85, 87, 91, 92, 94, 95, 194, 249
Maxwell, J.C. 225
McFetridge, I. 216
McLaughlin, B. 233
Melia, J. 41, 233
Mellor, D.H. 10, 222, 226, 229, 243
Menzel, C. 178
Merricks, T. 140
Mill, J.S. 43, 169, 170
Miller, K. 7, 141, 146, 156
Molnar, G. 97, 164, 165, 195
Mulligan, K. 72, 78, 95, 110, 207, 215
Mumford, S. 8, 20, 23, 24, 116, 118, 163, 165, 168, 171, 172, 193, 200, 201, 235

Newton, I. 223
Nolan, D. 189

O'Conaill, D. 208
Oliver, A. 15, 32, 226, 233
Orilia, F. 10, 238, 247–249

Paolini-Paoletti, M. 9
Parsons, J. 141, 146–148, 152, 156, 212
Pihlström, S. 215, 216
Place, U.T. 198
Plantinga, A. 105
Plato 23, 25, 28, 33, 55, 64, 87, 127

Quine, W.V.O. 34, 36–38, 40, 43, 44, 46, 204, 219, 220, 253

Ramsey, F.P. 46, 169, 219
Rissler, J. 21
Rodríguez-Pereyra, G. 32, 41, 48, 209, 212
Rosen, G. 189
Russell, B. 75, 77, 80, 81
Ryle, G. 161, 162, 222

Schaffer, J. 148, 234, 238, 239, 249
Schmaltz, T. 163
Schnieder, B. 46
Schrödinger, E. 78
Searle, J.R. 37
Shoemaker, S. 226
Sider, T. 139, 186, 187, 236
Simons, P. 6, 72, 74, 78–80, 82, 91, 95, 146, 207, 215
Skyrms B. 181
Smith, B. 72, 78, 95, 207, 215
Smolin, L. 82
Stoneham, T. 177
Stout, G.F. 86
Strawson, P.F. 43, 46, 110, 112, 228
Swinburne, R. 164, 167
Swoyer, C. 42

Tahko, T.E. 9, 208
Tarski, A. 220
Tegtmeier, E. 108, 127
Tooley, M. 96
Trogdon, K. 148
Tugby, M. 5, 27, 202, 205

Unger, R. M. 82

Vallicella, W.F. 7, 19, 105, 112, 122, 247
Valore, P. 10
Van Fraassen, B. 98
van Inwagen, P. 6, 48, 54, 56, 67, 140, 144

Watson, D. 177
Whitehead, A.N. 80
Wiggins, D. 43
Williams, D.C. 73, 75, 89
Wittgenstein L. 181, 186
Wolterstorff, N. 43

www.ingramcontent.com/pod-product-compliance
Lightning Source LLC
Chambersburg PA
CBHW050105170426
43198CB00014B/2465